REWARD

REWARD

Starter

Student's Book

Simon Greenall

MACMILLAN
HEINEMANN
English Language Teaching

Map of the book

Lesson	Grammar and functions	Vocabulary	Skills and sounds
1 *Hello, I'm Frank* Greeting people you don't know	Asking and saying names	*Hello, goodbye*	**Listening:** listening and matching; word recognition **Sounds:** pronouncing phrases **Reading:** inserting sentences **Speaking:** acting out a conversation
2 *I'm a student* Jobs	The indefinite article *a/an* Talking about jobs	Jobs	**Sounds:** stressing syllables in words **Listening:** listening and reading; ordering sentences **Reading:** reading and matching **Speaking:** acting out conversations
3 *How are you?* Greeting people you know	Greeting people Asking for and saying telephone numbers	Numbers 1-10	**Sounds:** pronouncing and recognising numbers **Listening:** listening and reading; ordering sentences **Reading:** reading telephone numbers **Speaking:** asking and saying telephone numbers
4 *Are you James Bond?* Asking and saying names	Asking and saying names Spelling	The alphabet	**Sounds:** pronouncing the alphabet; spelling names **Listening:** listening and repeating a conversation **Speaking:** acting out a conversation; finding out other people's professions
5 *She's Russian* Talking about where people are from	Saying where people are from Saying what nationality people are	Countries Nationalities	**Sounds:** pronouncing countries and nationalities **Listening:** listening and reading; listening and correcting wrong information **Reading:** reading and matching; completing charts **Writing:** writing sentences about people and their nationalities
6 *Is she married?* Asking and giving personal information	*Yes/no* questions and short answers	Numbers 11-20 Personal information	**Sounds:** pronouncing and recognising numbers **Listening:** listening and following a conversation; putting sentences in the right order **Writing:** copying and completing a membership form
7 *How old is he?* Talking about ages	Asking and saying how old people are Present simple (review)	Numbers 21-100	**Sounds:** pronouncing and recognising numbers **Listening:** listening and matching; listening and noting down personal information **Writing:** writing a poster of someone famous
8 *His favourite car is a Porsche* Talking about favourite people and things	*Who* and *what* Possessive adjectives: *my, your, his, her*	Favourite people and things	**Sounds:** pronouncing words; pronouncing questions **Listening:** listening and reading **Reading:** reading and answering a questionnaire **Listening:** listening and answering questions **Writing:** writing questions **Speaking:** talking about favourite people and things
9 *We're twins* Common interests	Present simple: *we're, you're, they're* Plurals	Words for people and their relationship to each other, eg *friend, twin, brother*	**Reading:** inserting sentences; reading and correcting wrong information **Listening:** listening and following a conversation
10 *What are these?* Finding out what the word for something is	Asking and saying what things are *This, that, these,* and *those*	Personal possessions Loan words	**Sounds:** pronouncing words; /æ/, /ɒ/, /ɪ/ **Listening:** listening and reading **Speaking:** asking and saying what things are in the classroom; playing *Word Bingo*
Progress check lessons 1-10	Revision	Word categories quiz	**Sounds:** /æ/, /e/, /ɔɪ/; stressing syllables in words; stressing words in sentences **Listening:** listening to a song - *Rock around the clock*

Lesson	Grammar and functions	Vocabulary	Skills and sounds
11 *How much are they?* The price of clothes and other items	Talking about prices Position of adjectives The definite article *the*	Clothes Money	**Sounds:** stressing words in sentences **Listening:** listening and correcting a conversation; listening for specific information to complete a chart **Speaking:** comparing prices
12 *Where are Jane's keys?* Personal possessions	Prepositions of place: *in, on, under* Possessive *'s*	Personal possessions Furniture *table, chair*	**Sounds:** pronouncing words **Listening:** listening and matching **Speaking:** asking and saying where things are **Writing:** writing sentences saying where things are in a picture
13 *We've got three children* Talking about families	*Have got* Possessive adjectives: *our, your, their*	Members of the family	**Reading:** reading and matching; completing a chart **Listening:** inserting sentences **Sounds:** /ə/ **Speaking:** talking about families; completing a chart
14 *She's got fair hair and blue eyes* Appearance and character	Talking about appearance and character *Has got*	Adjectives to describe appearance and character Colours Modifiers *quite, very*	**Speaking:** describing people **Listening:** listening and following a conversation **Reading:** completing charts; reading and matching **Writing:** writing about your family
15 *Stand up!* Instructions in the classroom	Imperatives	Features of a room Personal possessions	**Listening:** listening and following instructions **Sounds:** /əʊ/; stressing words in sentences; pronouncing instructions **Reading:** reading and checking comprehension **Writing:** writing instructions for a Students' Charter
16 *We live in a flat in Florence* Where people live and work	Present simple: regular verbs *I, we, you, they* Prepositions of place: *in, to*	Places where people live and work	**Sounds:** pronouncing words; /æ/, /əʊ/, /ɒ/, /uː/, /ɪ/, /ɜː/ **Reading:** reading and matching **Listening:** listening and checking **Reading:** reading and answering a game *What's my name?* **Writing:** writing your own *What's my name?*; writing about where you live
17 *What's the time?* Meal times in different countries	Telling the time (1) Present simple: *have* Prepositions of time: *at, in*	Meals Times of the day: hours	**Listening:** listening and reading **Reading:** reading and completing a chart **Speaking:** talking about when people around the world have meals
18 *I don't like Monday mornings* Weekly routines	Present simple: negatives Preposition of time: *on*	Days of the week Routine activities	**Sounds:** pronouncing days of the week; pronouncing expressions of time **Reading:** reading and matching **Listening:** listening and reading **Writing:** writing about weekly routines **Speaking:** talking about weekly routines
19 *Do you like running?* Likes and dislikes	*Yes/no* questions and short answers	Sports	**Sounds:** pronouncing words **Listening::** completing a chart; putting sentences in the right order **Writing:** completing a conversation **Speaking:** finding people who like and don't like sports
20 *She likes her job* Daily routines	Telling the time (2) Present simple: *he, she, it*	Times of the day: *quarter past/to, half past* Routine activities	**Speaking:** talking about routine activities; deducing information about someone **Reading:** reading and matching; reading and completing sentences **Listening:** listening and checking; listening for specific information **Writing:** writing a paragraph about two people's daily routines
Progress check lessons 11-20	Revision	Word categories quiz	**Sounds:** /ɑː/, /ʌ/, /ɜː/, /eə/; stressing syllables in words; stressing words in sentences **Listening:** listening to a song - *Three Little Birds*

Lesson	Grammar and functions	Vocabulary	Skills and sounds
21 *Does she go to work by boat?* Transport in cities	*He, she, it* *Yes/no* questions and short answers *By*	Means of transport	**Sounds:** pronouncing words and sentences; /eɪ/, /əʊ/, /ɑː/, /aɪ/, /ʌ/, /ɔː/, /æ/ **Reading:** reading and completing a chart **Listening:** listening and completing a chart **Speaking:** acting out conversations; completing a questionnaire on how people get to work
22 *What do they eat in Morocco?* Food and drink in different countries	Present simple: *Wh*-questions	Food and drink Food from different countries	**Reading:** reading and matching; completing a chart **Listening:** listening and checking **Writing:** writing a paragraph about food and drink **Speaking:** talking about food and drink
23 *I don't like lying on the beach* Holiday activities	*Like + ing* Present simple: negatives	Holiday activities	**Reading:** reading and checking comprehension **Listening:** listening and completing a chart **Speaking:** asking and answering questions about holidays **Writing:** writing about your partner's holiday
24 *There's a telephone in the hall* Homes	*There is/are* *Any*	Rooms of the house Furniture	**Sounds:** pronouncing words **Reading:** reading and checking comprehension **Listening:** listening to a commentary; listening and deducing **Speaking:** describing a room in your house **Writing:** writing a description of a room in your house
25 *I usually have a party* Birthday celebrations	Present simple: adverbs of frequency	Months of the year Ordinal numbers	**Sounds:** pronouncing words **Reading:** reading and answering questions **Listening:** listening and completing a chart **Speaking:** asking and answering questions about birthdays
26 *I can cook* Abilities for jobs	*Can* for ability	Abilities eg *cook, draw, drive, sing*	**Sounds:** pronouncing the strong and weak form of *can* **Reading:** reading and making notes **Listening:** listening for specific information **Speaking:** acting out a job interview
27 *Can I have a sandwich, please?* Restaurant menus	Talking about food and drink	Food and drink	**Sounds:** pronouncing expressions such as *a cup of tea, a glass of wine* **Listening:** listening and inserting sentences; listening for specific information **Speaking:** acting out conversations in a restaurant
28 *Where's the station?* Finding your way around town	Asking for and giving directions	Shops Directions, eg *go straight ahead, turn left*	**Sounds:** pronouncing words; stressing syllables in words **Listening:** listening and matching; following a route on a map **Reading:** reading and following a route on a map **Writing:** writing a guided tour of your city
29 *He's buying lunch* Times and actions around the world	Present continuous (1)	Words for actions such as *buy, drive, sit, run*	**Reading:** reading and matching **Sounds:** pronouncing the present continuous **Speaking:** describing what people are doing in a picture and around the world
30 *He isn't having a bath* Talking about what's happening at the moment	Present continuous (2): negatives; questions	Nouns which go with certain verbs, *have, make*	**Reading:** reading and matching **Listening:** listening for specific information **Speaking:** saying the differences between two pictures **Writing:** writing the differences between two pictures
Progress check lessons 21-30	Revision	Identifying parts of speech Verbs and nouns which go together Word categories quiz	**Sounds:** /ɔɪ/, /ɪ/, /uː/, /iː/, /ʊ/; stressing syllables in words; stressing words in sentences: contrastive stress **Listening:** listening to a song - *Daniel*

Lesson	Grammar and functions	Vocabulary	Skills and sounds
31 *We're going to Australia* A visit to Australia	Present continuous (3): future plans	Words for travel	**Reading:** reading and answering questions **Listening:** listening and checking; listening for specific information **Speaking:** planning a trip to somewhere special
32 *Let's go to the cinema* Entertainment	Making suggestions Accepting and refusing Talking about the cinema and theatre	Places and forms of entertainment	**Listening:** listening and checking **Sounds:** using suitable intonation; stressing words in sentences **Speaking:** making, accepting and refusing suggestions to do things in your town
33 *Yesterday, I was in Paris* Weekends	Past simple (1) *be*: *was/were* affirmative Expressions of time	Adjectives to describe moods and feelings, such as *happy, cold*	**Listening:** listening and reading **Sounds:** pronouncing the strong and weak forms of *was* and *were*; pronouncing sentences **Reading:** reading and answering questions **Writing:** completing a passage
34 *Was she in the kitchen?* A murder mystery	Past simple (2): *yes/no* questions and short answers	New words from the murder mystery	**Reading:** predicting, reading and matching **Listening:** listening and checking; completing a chart **Speaking:** talking about the murder mystery
35 *They didn't have any computers* Technology and household goods in the past	Past simple (3): *had*	Household equipment	**Sounds:** stressing syllables in words **Reading:** reading and completing a chart **Listening:** listening for specific information **Speaking:** talking about an anachronistic picture; talking about inventions and discoveries
36 *We listened to the radio* Childhood	Past simple (4): regular verbs	Expressions of past time	**Reading:** reading and matching; correcting sentences **Sounds:** pronouncing the endings of regular verbs **Listening:** listening for specific information **Writing:** writing sentences about yourself ten years ago
37 *Picasso didn't live in Spain* The life of Pablo Picasso	Past simple (5): negatives	Verbs to describe the life of Picasso	**Reading:** predicting; reading and checking **Sounds:** stressing words in sentences: contrastive stress **Speaking:** talking about the life of a famous person **Writing:** writing about the life of a famous person
38 *Did you take a photograph?* Animal mysteries	Past simple (6): *yes/no* questions and short answers	Animals	**Reading:** inserting sentences **Listening:** listening and checking **Sounds:** using suitable intonation for questions **Reading:** reading and answering questions; predicting a story
39 *We went to New York* A visit to New York	Past simple (7): questions; irregular verbs	Paper documents, such as *bills, ticket, receipts*	**Sounds:** pronouncing words **Listening:** listening and reading **Reading:** reading and answering questions **Writing:** completing a letter; writing about a visit **Speaking:** checking information
40 *The end of the world?* A short story	Tense review: present simple, present continuous, past simple	New words from *The end of the world?*	**Reading:** reading and answering questions **Listening:** listening and checking **Speaking:** predicting the end of the story
Progress check lessons 31–40	Revision	Different ways of recording vocabulary Survival vocabulary Word categories quiz	**Sounds:** /ɪ/, /iː/, /ɑ/, /ʊ/and /ɔː/ **Listening:** listening to a song – *Yesterday*

Classroom language

Listen

Look

Say

Write

Read

Underline

Tick

Check

Punctuate

Spell

Answer

Work in pairs

Complete

Ask

Copy

Repeat

Correct

Point

Act

Count

Match

Number

Put in order

Turn to page 3

Translate

1 *Hello, I'm Frank*

Asking and saying names

LISTENING AND READING

1 Match the conversations with the photos.

> **A FRANK** Hello, I'm Frank.
> What's your name?
>
> **SARAH** Hello, Frank.
> I'm Sarah.
>
> **B KATE** Goodbye Pete!
>
> **MIKE** Goodbye!

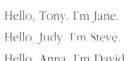 Now listen and read.

2 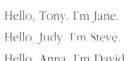 Listen and match.

> Anna   David
>
> Tony   Steve
>
> Judy   Jane

3 Read and match.

> Hello, I'm Anna. What's your name?   Hello, Tony. I'm Jane.
>
> Hello, I'm Tony. What's your name?   Hello, Judy. I'm Steve.
>
> Hello, I'm Judy. What's your name?   Hello, Anna. I'm David.

 Now listen again and check.

VOCABULARY AND SOUNDS

1 Listen and repeat.

> hello   goodbye

2 Listen and repeat.

> name   your name   What's your name?
>
> Hello, what's your name?
>
> Frank   I'm Frank.
>
> Hello, what's your name?
>
> Judy   I'm Judy.   Hello, I'm Judy.
>
> Goodbye, Judy!

FUNCTIONS

> **Asking and saying names**
> *What's your name?*
> *(= What is your name?)*
> *I'm Kate. (= I am Kate.)*

1 Complete.

1 Hello. What's ____ name?

2 ____ John.

3 ____ your name?

4 Hello. I ____ Petra.

1 Hello. What's your name?

2 Write and punctuate.

1 whats your name

2 im sarah

3 hello im gerry whats your name

4 hello gerry im henry

3 Match the sentences with the speakers in the photos above.

a Hello, Val, I'm Katy.

b Hello, Katy.

c Hello, I'm Val. What's your name?

4 Ask and say your names.

Hello, I'm ____. What's your name?

Hello, ____. I'm ____.

READING AND SPEAKING

1 Read and complete.

STEVE	Hello. What's your name?	a	Hello, Steve.
SEMA	(1) ____	b	What's your name?
STEVE	Hello, Sema.	c	I'm Sema.
SEMA	(2) ____		
STEVE	I'm Steve.		
SEMA	(3) ____		

2 Work in pairs. Act out the conversation in 1.

3 Find three more words and phrases from this lesson.

X	C	B	N	K	N	H	M	N	H	K	P	O
G	H	J	K	L	G	E	B	K	J	L	A	S
G	G	Q	F	R	S	L	D	Z	X	V	K	L
N	M	K	L	T	H	L	D	F	G	V	I	F
W	H	A	T	S	Y	O	U	R	N	A	M	E
D	Q	B	O	Y	F	D	S	H	J	A	K	C
Y	I	O	P	B	F	G	H	K	D	F	A	U
Q	S	D	G	H	C	V	N	B	G	J	T	L
A	X	C	V	G	B	G	O	O	D	B	Y	E

4 Say goodbye to people at the end of the lesson.

Goodbye, Yildiz.   Goodbye, Frank.

2 | *I'm a student*

The indefinite article *a/an*; talking about jobs

1   2   3   4   5

6   7   8   9   10

VOCABULARY AND SOUNDS

1 🔊 Listen and repeat.

> doctor   teacher   secretary   actor
> journalist   singer   student   waiter
> actress   engineer   job

2 🔊 Listen and match.

Pete   a teacher
José   b doctor
Hillary   c journalist
Maria   d student
Yıldız   e waiter
Hashimi   f singer
                g secretary
                h actor
                i engineer
                j actress

Pete – c

Now write the four extra jobs.

3 🔊 Listen and repeat.

job   what's   What's your job?
doctor   a doctor   I'm a doctor.
engineer   an engineer
I'm an engineer.

LISTENING

1 🔊 Listen and read.

JOSÉ What's your job, Pete?

PETÉ I'm a journalist. What's your job, José?

JOSÉ I'm a doctor.

2 Number the sentences in the right order.

☐ Hillary   I'm a secretary. What's your job, Maria?

☐ Maria   I'm a teacher.

☐ Maria   What's your job, Hillary?

🔊 Now listen and check.

GRAMMAR AND FUNCTIONS

The indefinite article *a* and *an*

I'm a waiter.   NOT ~~I'm waiter.~~

I'm a doctor.   NOT ~~I'm doctor.~~

I'm an actor.   NOT ~~I'm a actor.~~

I'm an engineer.   NOT ~~I'm a engineer.~~

Talking about jobs

What's your job?   *I'm a waiter.*   (= I am)

                I'm a doctor.

                I'm an actor.

                I'm an engineer.

---

1 Write *a* or *an*.

1 I'm ____ doctor.   4 I'm ____ actor.

2 I'm ____ student.   5 I'm ____ journalist.

3 I'm ____ engineer. 6 I'm ____ waiter.

2 Write *'s* or *'m*.

1 What ____ your job?   3 I ____ a doctor.

2 What ____ your name?   4 My name ____ Bob.

3 Complete.

Hashimi   What's ____ job, Yıldız?

Yıldız   I'm a student. ____ your job, Hashimi?

Hashimi   I'm ____ engineer.

4 Answer the questions.

1 What's your name?

2 What's your job?

READING AND SPEAKING

1 Match the sentences with the photos.

a Hello, I'm Jacqueline. I'm a teacher.

b Hello, I'm Philip. I'm a doctor.

c Hello, I'm Angie. I'm a student.

d Hello, I'm Mel. I'm a waiter.

2 The sentences in 1 are true. Correct the conversations below.

Hello, what's your name?

Hello, I'm Philip. What's your name?

I'm Mel. What's your job?

I'm a student. What's your job, Mel?

I'm a doctor.

Hello, what's your name?

Hello, I'm Angie. What's your name?

I'm Jacqueline. What's your job?

I'm a teacher. What's your job, Jacqueline?

I'm a doctor.

3 Work in pairs and act out the correct conversations.

4 Work in pairs and act out your own conversations.

3 How are you?

Greeting people; asking for and saying telephone numbers

VOCABULARY AND SOUNDS

1 🔈 Listen and repeat.

0 zero (oh)	1 one	2 two
3 three	4 four	5 five	6 six
7 seven	8 eight	9 nine	10 ten

2 Say.

2 6 9 3 1 8 4
7 10 5 6 3 1

🔈 Now listen and check.

3 Complete.

tw_ o_ _ s_ _ ei_ _ _ th_ _ _
se_ _ _ fi_ _ fo_ _ te_ n_ _ _

4 Work in pairs. Say and write numbers.

Student A: (Say) *two*
Student B: (Write) *2*

5 🔈 Listen to these telephone numbers.

(tele)phone number

01223 589933   0171 454523
00 44 1 865 774356   01227 654398

6 Say these telephone numbers.

0175 348900   01456 88466
0378 015030   0161 778 4598

🔈 Now listen and check. As you listen, repeat the numbers.

LISTENING AND READING

1 🔈 Listen and read the conversations.

A FRANK Hello, Sarah. How are you?
SARAH   Hello, Frank. I'm fine, thanks. How are you?
FRANK   I'm very well, thank you.

B MIKE What's your telephone number, Sue?
SUE   0171 589 3245. What's your telephone number, Mike?
MIKE   01993 453298.

2 Number the sentences in the right order.

A ☐ I'm fine, thanks.
   ☐ Hello, Michiko. I'm very well, thanks. How are you?
   ☐ Hello, Joan. How are you?

B ☐ 01967 328123.
   ☐ Goodbye.
   ☐ 0134 521 3987. What's your telephone number, Pete?
   ☐ Kate, what's your telephone number, please?
   ☐ Thank you. Goodbye, Pete.

🔈 Now listen and check.

FUNCTIONS

> **Greeting people**
> *How are you? I'm fine, thanks. (= I am fine.)*
> *How are you? I'm very well, thank you. (= I am very well.)*
>
> **Asking for and saying telephone numbers**
> *What's your telephone number, please?*
> *Oh one seven one eight three seven seven two.*
> **Remember! You say *oh* for *0* in telephone numbers.**

1 Put the words in the right order and make sentences.

1 are you how hello?

2 you fine thank I'm

3 thanks very I'm well

4 your what's number please telephone ?

2 Read and complete.

a I'm very well, thank you.
b Hello, Val. How are you?

Hello, Henry.

(1) ____

I'm fine, thanks. How are you?

(2) ____

3 Complete.

____, Pete, how are ____?
Hello, Geoff. I'm ____, thanks. How ____ you?
____ very well, ____ you.

4 Work in pairs. Ask and say.

Hello, (name), how are you?
Hello, (name). I'm fine, thanks. How are you?
I'm very well, thank you.

5 Write in words.

0123 3345   01778 3456

01786 54332   01223 56765

Oh one two three, three three four five

READING AND SPEAKING

1 Read and say the numbers.

NAME	TELEPHONE NUMBER
Tim Clark	*0161 524 3345*
Will Bush	*01557 345877*
Dan Ford	*01287 890992*
Anna Green	*0181 227 4567*
Graham White	*01821 778 2695*
Jane Smith	*00 44 1873 456789*

2 Work in pairs.

Student A: Say a number.

Student B: Say the name.

Student A: 0181 227 4567.
Student B: Anna Green.

3 Ask and answer with other students.
Complete the chart.

What's your name?
What's your job?
What's your telephone number?

Name	Job	Telephone number
Bertrand	*Student*	*91 23 78 45 67*

Are you James Bond?

Asking and saying names; spelling

VOCABULARY AND SOUNDS

1 🔊 Listen and repeat.

| a b c d e f g h i j k l m n |
| o p q r s t u v w x y z |
| |
| A B C D E F G H I J K L M N |
| O P Q R S T U V W X Y Z |

2 🔊 Listen and repeat.

| /eɪ/ | /iː/ | /e/ | /aɪ/ |
| A H J K | B C D E G P T V | F L M N X Z | I |

| /əʊ/ | /uː/ | /ɑː/ |
| O | Q U | R |

3 🔊 Listen and repeat.

spell   how   How do you spell ... ?

How do you spell James?   How do you spell Bond?

How do you spell your name?

4 Work in pairs. Ask and say how you spell the names.

Francesca   Yıldız   Keiko   Xavier   Maria

How do you spell Francesca?   F-R-A-N-C-E-S-C-A

5 Work in pairs.

Student A: Turn to Communication activity 1 on page 92.

Student B: Turn to Communication activity 15 on page 94.

LISTENING AND SPEAKING

1 🔊 Listen and read.

MAN 1	Are you Count Dracula?
MAN 2	Yes, I am.
MAN 1	Thank you. Are you Cleopatra?
WOMAN 1	Yes, I am.
MAN 1	Thank you. Are you Sylvester Stallone?
MAN 3	No, I'm not.
MAN 1	Oh! Are you Frank Sinatra?
MAN 3	No, I'm not. I'm Bond – James Bond.
MAN 1	Ah ha! Mr Bond, thank you very much. Are you Frankenstein?
WOMAN 2	No, I'm not. I'm Morticia.
MAN 1	Morticia. How do you spell Morticia, please?
WOMAN 2	M-O-R-T-I-C-I-A.
MAN 1	Ah, yes, Morticia. Thank you very much.
WOMAN 2	Thank you.

2 Work in pairs. Act out the conversation in 1.

FUNCTIONS

> **Asking and saying names**
> *Are you Count Dracula?   Yes, I am.   NOT ~~Yes, I'm~~*
> *Are you Frank Sinatra?   No, I'm not. (= No, I am not.)*
>
> **Spelling**
> *How do you spell your name?   M-O-R-T-I-C-I-A.*

1 Complete.

Are ___ Philip?

Yes, I ____. Are you Michael?

No, ____ not.

What___ your ____?

My ____ Oonagh.

How do you ____ Oonagh?

O-O-N-A-G-H.

2 Put the words in the right order and make sentences.

1 you Yildiz are?

2 I'm no not.

3 your name what's?

4 Chris my name's.

5 you are Thomas?

6 yes am I.

3 Work in pairs. Ask and say.

Are you Marco?

No, I'm not. I'm (name). Are you (name)?

Yes, I am.

How do you spell (name)?

(Spell your name).

LISTENING AND SPEAKING

1 📼 Listen and take dictation.

Now match the names and the photos.

2 Choose a character. Work in pairs and act out the conversation.

Are you Rambo?   *No, I'm not.*

Are you James Bond?   *Yes, I am.*

3 Work in groups of four.

Student A: Turn to Communication activity 2 on page 92.

Student B: Turn to Communication activity 16 on page 95.

Student C: Turn to Communication activity 26 on page 97.

Student D: Turn to Communication activity 40 on page 100.

4 Work together and ask questions to find out who the other students are and what their jobs are.

Name	Henry Schwarzkopf Fiona Pink Dave Dingle Mike Handy Fifi Lamour Tom James Frank Fearless Adam Hackett
Job	actress doctor teacher singer student actor journalist engineer

Are you Henry Schwarzkopf?   *No, I'm not.*

Are you an actress?   *Yes, I am.*

The first person to find out all the names and the jobs is the winner.

5 She's Russian

Saying where people are from; saying what nationality people are

VOCABULARY AND SOUNDS

1 Match the correct countries with the photos.

> the United States of America   Britain   Turkey
> Italy   Brazil   Russia   Japan   Thailand

2 🔲 Listen and repeat.

Thailand   the United States of America
Britain   Turkey   Italy   Brazil   Russia   Japan

3 Count the syllables.

☐ ☐   ☐ ☐   ☐☐☐   ☐☐ ☐ ☐
Thailand   Japan   Italy   America

☐ ☐   ☐ ☐   ☐ ☐   ☐ ☐
Britain   Brazil   Turkey   Russia

🔲 Now listen and repeat.

4 Read.

> country   nationality

Britain is a *country*.   British is a *nationality*.

5 Match *country* and *nationality*.

> American   Japanese   Italian   Brazilian
> Thai   British   Turkish   Russian

American – the USA

🔲 Now listen and check. As you listen, repeat the words.

6 Write your country and nationality.

LISTENING AND READING

1 🔲 Listen and read.

A Steve Forrest is a doctor. He's British and he's from London.

B Henry Fuller is American. He's a waiter and he's from New York.

C Silvia Soares is from Rio. She's Brazilian and she's a student.

2 Complete.

Name	Steve Forrest				
Nationality					
Job					
From					

3 [cassette] Listen and underline anything which is different.

'Hello, I'm Olga Maintz. I'm a secretary. I'm Russian and I'm from Moscow.'

'Hi! I'm Mustafa Polat. I'm Turkish and I'm an actor. I'm from Ankara.'

'Hello! I'm Patrizio Giuliani. I'm from Rome. I'm Italian and I'm a teacher.'

4 Complete the chart with the information you hear in 3.

FUNCTIONS

Saying where people are from
I'm from Bangkok. (= I am)
He's from London. (= he is)
You're from Rome. (= you are)
She's from Rio. (= she is)

Saying what nationality people are
I'm Thai.   *He's British.*
You're Italian.   *She's Brazilian.*

1 Put the words in the right order and make sentences.

1 from Paris he's French and he's

2 Italian she's from Rome she's and

3 he's and teacher Thai he's a

4 I'm and actor from London an I'm

5 a Russian and I'm doctor I'm

6 you're from a you're secretary and New York

2 Write your answers to *Listening and reading* activity 3.

Olga Maintz is a
secretary. She's ...

3 Work in pairs.

Student A: Say where you're from.

Student B: Say what nationality Student A is.

I'm from Rome.   You're Italian.

WRITING AND SPEAKING

1 Complete the chart and write sentences.

Person	Nationality	Country
Pele	Brazilian	Brazil
Diana, Princess of Wales		
Pavarotti		
Cher		

Pele is Brazilian. He's from Brazil.

2 Work in groups of three.

Student A: Turn to Communication activity 3 on page 92.

Student B: Turn to Communication activity 10 on page 93.

Student C: Turn to Communication activity 22 on page 96.

3 Work together and complete the chart.

Name	Nationality	Job	From

6 Is she married?

Yes/no questions and short answers

VOCABULARY AND SOUNDS

1 🔊 Listen and repeat.

11	eleven	16	sixteen
12	twelve	17	seventeen
13	thirteen	18	eighteen
14	fourteen	19	nineteen
15	fifteen	20	twenty

2 Say.

11 14 18 13 20 19 15
16 11 17 12 18 20

🔊 Now listen and check.

3 🔊 Listen and tick.

1 ☐	6 ☐	11 ☐	16 ☐
2 ☐	7 ☐	12 ☐	17 ☐
3 ☐	8 ☐	13 ☐	18 ☐
4 ☐	9 ☐	14 ☐	19 ☐
5 ☐	10 ☐	15 ☐	20 ☐

4 Now play *Numbers Bingo*.

> **Numbers Bingo**
>
> 1 Complete the chart with numbers from 1 – 20
> 2 Work in groups of four or five. One of you says numbers from 1 – 20.
> 3 Tick (✔) the numbers in your chart if they are there.
> 4 Are there five ticks (✔) in a line? Yes? Say *Bingo*!

5		13		8
	7			
			12	

5 Complete Shirley Smith's card with these words.

name married nationality job address age

LISTENING AND WRITING

1 🔊 Listen and read.

Is Shirley Smith from Kenton?

Yes, she is.

Is she married?

No, she isn't. She's sixteen.

Is she a student?

Yes, she is.

And she's British?

Yes.

2 Number the sentences in the right order.

☐ And is he British?

☐ No, he isn't. He's American.

☐ Is Ken Stanwell from Kenton?

☐ Yes, he is.

☐ Is he married?

☐ Yes, he is.

☐ No, he isn't. He's seventeen.

☐ Is he a student?

🔊 Now listen and check.

3 Copy and complete the Membership card for Ken Stanwell.

GRAMMAR

> **Yes/no questions and short answers**
>
> | | | |
> |---|---|---|
> | *Are you Italian?* | *Yes, I am.* | *No, I'm not.* |
> | *Is he from Rome?* | *Yes, he is.* | *No, he isn't.* |
> | *Is she married?* | *Yes, she is.* | *No, she isn't.* |
> | *Is your name Dave?* | *Yes, it is.* | *No, it isn't.* |

1 Match the questions and answers.

1 Are you English?   a Yes, he is.

2 Is Ken Stanwell from Kenton?   b Yes, she is.

3 Is Ken Stanwell married?   c No, she isn't.

4 Is Shirley Smith from Kenton?   d Yes, it is.

5 Is Shirley Smith married?   e No, I'm not.

6 Is a Fiat Panda Italian?   f No, he isn't.

7 Is a Mercedes Brazilian?   g No, it isn't.

2 Work in pairs and act out the conversations in *Listening and writing* activities 1 and 2.

3 🔊 Listen and put a tick (✔) or a cross (✗).

	Jane	Anna	Sema	Kazuo	Steve
Married	✔	✗			

4 Match the questions and answers.

1 Is Jane married?   a Yes, she is.

2 Is Kazuo married?   b No, she isn't.

3 Is Anna married?   c Yes, he is.

4 Is Steve married?   d No, she isn't.

5 Is Sema married?

5 Work in pairs. Check 4.

Is Jane married?   *Yes, she is.*
Is Anna married?   *No, she isn't.*

6 Tick (✔) the correct sentence.

1 Are you married?   a Yes, I'm.   b Yes, I am.

2 Is she a secretary?   a Yes, she is.   b Yes, she's.

3 Is he from Rio?   a Yes, he is.   b Yes, it is.

4 a Is she Japanese?   No, she isn't.
  b Are she Japanese?   No, she isn't.

SPEAKING AND LISTENING

1 Work in pairs. Write six questions about another student and ask and answer.

Is Kazuo Japanese?   *Yes, he is./No, he isn't.*

Now check your answers.

Are you Japanese, Kazuo?   *Yes, I am.*

2 Work in pairs. Ask and answer.

> **PLAY 20 QUESTIONS!**
> 1 Is Bill Clinton an engineer?
> 2 Is Tom Cruise an actor?
> 3 Is pizza from Italy?
> 4 Is 'doctor' a job?
> 5 Is San Francisco in the United States?
> 6 Are you from Japan?
> 7 Is 'Graham' a French name?
> 8 Is Istanbul a country?
> 9 Is seventeen 17?
> 10 Is Whitney Houston American?
> 11 Are you President of the USA?
> 12 Is 'Argentinian' a nationality?
> 13 Is Edinburgh in England?
> 14 Is Spain a country?
> 15 Is Roberto Baggio an actor?
> 16 Are you André Agassi?
> 17 Is Sony Korean?
> 18 Is champagne from France?
> 19 Is your name Queen Elizabeth?
> 20 Are you married?

🔊 Now listen and check.

REGISTRATION OF MARRIAGES IN NORTHERN IRELAND

Certified Copy of an Entry of Marriage

7 & 8 Vict., Cap. 81; 26 Vict., Cap. 27

North Down

solemnized at Registrar Office In the District of North D...

Name and surname	Age	Condition	Rank or profession	Residence at the ti...
Timothy Eoin Friers	15 May 1962 / 33 years	Bachelor	Graphic Designer	33, Victoria 15 Holywo...
	10 February	Spinster	Graphic Designer	23, Napier Coyle...

7 How old is he?

Asking and saying how old people are; present simple (review)

VOCABULARY AND SOUNDS

1 📼 Listen and repeat.

21	twenty-one	30	thirty
22	twenty-two	31	thirty-one
23	twenty-three	40	forty
24	twenty-four	50	fifty
25	twenty-five	60	sixty
26	twenty-six	70	seventy
27	twenty-seven	80	eighty
28	twenty-eight	90	ninety
29	twenty-nine	100	one hundred

2 📼 Listen and repeat.

☐☐ thirteen   fourteen   fifteen   sixteen
☐☐☐ seventeen
☐☐ eighteen   nineteen
☐☐ thirty   forty   fifty   sixty
☐☐☐ seventy
☐☐ eighty   ninety

3 📼 Listen and tick.

13 ☐	30 ☐
14 ☐	40 ☐
15 ☐	50 ☐
16 ☐	60 ☐
17 ☐	70 ☐
18 ☐	80 ☐
19 ☐	90 ☐

4 Match.

15	ninety-three
67	fifteen
93	forty-five
45	fifty-four
32	eighty-eight
88	sixty-seven
54	thirty-two

5 Work in pairs.

Student A: Say a number.

Student B: Point to the photo.

LISTENING

1 📼 Listen and match.

1	Tony	19
2	Karen	27
3	Nick	20
4	Sarah	17
5	Jill	23
6	Alex	35

2 Work in pairs and check.

How old is Tony?   He's twenty.

GRAMMAR AND FUNCTIONS

Asking and saying how old people are	
How old are you?	*I'm thirty-five.*
	I'm thirty-five years old.
How old is he?	*He's nineteen.*
How old is she?	*She's twenty-seven.*

Present simple (review)

Affirmative	Negatives	Questions
I'm	*I'm not*	*Am I?*
you're	*you aren't*	*Are you?*
he's	*he isn't*	*Is he?*
she's	*she isn't*	*Is she?*
it's	*it isn't*	*Is it?*

1 Punctuate and write questions and answers.

1 how old are you

2 how old is she

3 how old is he

4 shes twentythree

5 hes thirtyfive

6 im twentytwo.

1 How old are you?

2 Match the questions and answers in 1.

How old are you?   I'm twenty-two.

3 Match the questions and answers.

1	Is Tony fifty-two?	a	Yes, he is.
2	Is Karen thirty-five?	b	No, she isn't. She's twenty-three.
3	Is Nick nineteen?	c	No, she isn't. She's thirty-five.
4	Is Sarah nineteen?	d	No, she isn't. She's twenty-seven.
5	Is Jill eighty-nine?	e	No, he isn't. He's twenty.
6	Is Alex seventeen?		

4 Work in pairs. Check 3.

Is Tony fifty-two?   No, he isn't.

5 Ask and answer about the ages of students in the class.

Is Marco thirty?   Yes, he is./No, he isn't.

LISTENING AND WRITING

1 🔊 Listen and match the name with the picture.

Ella   Miki   Maria   Erol   Carlos   Anant

1 – Ella

2 Work in pairs.

Student A: Talk about the people in the pictures.

Student B: Point to the correct picture.

She isn't married. She's 42.
She's an actress. She's American.

3 Make a poster of someone famous. Write sentences about:

– job

– where he or she's from

– nationality

– if he or she is married

– how old he or she is

4 Look at the vocabulary boxes in lessons 1–7. Count the words you know.

5 Work in pairs and check your answers.

Student A: Say words you know in lessons 1–7.

Student B: Translate the words Student A says.

8 | *His favourite car is a Porsche*

Who and **what**; possessive adjectives: *my, your, his, her*

VOCABULARY AND SOUNDS

1 Match the words with the photos.

> group   politician   football team
> TV programme   TV presenter   car

1 car

2 🔊 Listen and check. As you listen, repeat the words.

3 🔊 Listen and repeat.

> favourite

singer   favourite   your favourite singer
Who's your favourite singer?

car   favourite   your favourite car
What's your favourite car?

actor   favourite   your favourite actor
Who's your favourite actor?

group   favourite   your favourite group
What's your favourite group?

READING AND LISTENING

1 Read and answer *Favourite people ... Favourite things*.

Favourite people ... Favourite things	
Who's your favourite	singer?
	actor?
	politician?
	actress?
What's your favourite	car?
	group?
	football team?
	TV programme?

2 🔊 Read and listen.

Samantha Alton is a secretary. She's twenty and she's from Birmingham. Her favourite singer is Paul Young and her favourite actor is Sylvester Stallone. Her favourite group is U2 and her favourite TV programme is The Clothes Show.

Bill Henderson is a student. He's American and he's from Los Angeles. His favourite politician is Bill Clinton and his favourite actress is Sharon Stone. His favourite car is a Mercedes and his favourite American football team is the Chicago Bears.

3 Write the questions to Samantha and Bill's answers.

Samantha: Who's your favourite singer/actor?

What's your favourite ...?

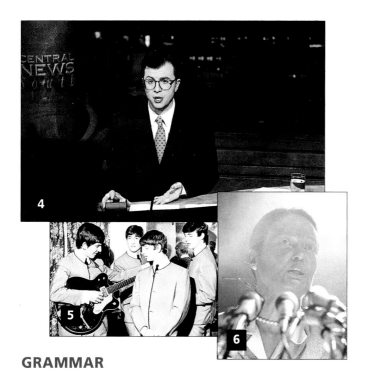

GRAMMAR

> ### Who and what
>
> *Who's your favourite singer?*
> *Who's your favourite teacher?*
> *What's your favourite car?*
> *What's your favourite TV programme?*
>
> ### Possessive adjectives
>
> | | my | | My | |
> |---|---|---|---|---|
> | *What's* | *your* | *name?* | *Your* | *name's Pat.* |
> | | his | | His | |
> | | her | | Her | |

1 Complete with *who* or *what*.

1 _____ is your teacher?

2 _____ is your name?

3 _____ is your favourite country?

4 _____ is your address?

5 _____ is fifteen?

6 _____ is your job?

2 Complete.

I	my
you	
he	
she	

3 Write questions using *his/her*.

Samantha's favourite singer
Who's her favourite singer?

Bill's favourite car
What's his favourite car?

1 Bill's favourite American football team

_____

2 Samantha's favourite actor

_____

3 Samantha's favourite group

_____

4 Bill's favourite actress

_____

4 Write full answers to the questions in 3.

Samantha: Her favourite singer is Paul Young.
Bill: His favourite car is a Mercedes.

WRITING AND LISTENING

1 Here are some answers. Write the questions.

1 a Porsche ☐
2 Arnold Schwarzenegger ☐
3 Manchester United ☐
4 the Beatles ☐
5 Diana Ross ☐

1 What's your favourite car?

2 📼 Listen to Max and Sally talking about their favourite people and things. Write M for Max and S for Sally by the answers in 1.

3 Work in pairs and check your answers.

His favourite car is a Porsche.
Her favourite actor is ...

4 Work in pairs. Talk about your favourite people and favourite things.

Who's your favourite singer?   Diana Ross.

5 Tell the rest of the class about your partner's favourite people and things.

His/Her favourite singer is Diana Ross.

9 | We're twins

Present simple: we're, you're, they're; plurals

VOCABULARY AND SOUNDS

1 Match the sentences with the photos.

friend twin neighbour

a 'We're friends.'
b 'We're neighbours.'
c 'We're twins.'

2 Match singular and plural words.

brother sister boy girl man woman

sisters boys girls women brothers men

3 Match the sentences with the photos in 1.

a 'They're girls.'   b 'They're men.'
c 'They're women.'   d 'They're brothers.'

4 🔲 Listen and repeat.

/z/ friends neighbours boys girls brothers
sisters twins

READING AND LISTENING

1 Read and complete.

JANE	Are you twins?
NICK	Yes we are. I'm Nick.
DAVE	And I'm Dave.
JANE	Are you from London?
NICK/DAVE	(1) _____ . We're from Manchester.
JANE	How old are you?
NICK/DAVE	(2) _____ .
JANE	What are your jobs?
NICK/DAVE	(3) _____ .
JANE	What's your favourite football team?
NICK/DAVE	(4) _____ .

a We're twenty-three.
b Manchester United.
c No, we aren't.
d We're students.

2 Work in pairs and check your answers.
🔲 Now listen and check.

3 Read the conversation and correct any information which is wrong.

PAUL	Who are they?
JANE	They're ... er Steve and Dick.
PAUL	Are they from London?
JANE	No, they're from, er, ... Liverpool.
PAUL	And how old are they?
JANE	They're eighteen. They're actors.

🔲 Now listen and check.

GRAMMAR

Present simple: *we're, you're, they're*		
we're (= *we are*)	*we aren't*	(= *we are not*)
you're (= *you are*)	*you aren't*	(= *you are not*)
they're (= *they are*)	*they aren't*	(= *they are not*)

Plurals

Singular	Plural
brother —	*brother***s**
sister	*sister***s**
boy	*boy***s**
girl	*girl***s**
BUT	
woman	*women*
man	*men*

1 Complete with '*m*, '*s* or '*re*.

1 We ____ doctors.

2 They ____ from Manchester.

3 I ____ British.

4 He ____ a student.

5 We ____ fine, thank you.

6 They ____ French.

2 Correct the information in *Reading and listening* activity 3.

They aren't Steve and Dick.

They're Dave and Nick.

3 Work in pairs. Write true sentences with '*re* or *aren't* .

1 We ____ students. 4 We ____ seventeen.

2 We ____ British. 5 We ____ friends.

3 We ____ from Bangkok. 6 We ____ neighbours.

4 Work with another pair and tell them your answers to 3. Then tell the class about each other.

They're students.

They aren't British.

5 Answer the question.

How do you form most plurals?

6 Tick (✔) the correct phrase or sentence.

1 a Two student. b Two students.

2 a Three books. b Three book.

3 a A football team. b A football teams.

4 a My favourite car is a Porsche.
  b My favourite cars is a Porsche.

5 a They're doctor. b They're doctors.

6 a We're neighbour. b We're neighbours.

READING AND WRITING

1 Read and complete the chart.

'We're Kasem and Ladda. We're friends. We're Thai and we're from Chiang Mai. We aren't married. We're seventeen and sixteen.'

Names	Kasem and Ladda	Students A and B
Nationality		
From		
Age		

2 Work in pairs. Complete the chart.

We're Kemal and Erol.

We're Turkish.

3 Find someone you have a lot in common with. Tell the class.

We're sixteen. We're from Italy...

10 | *What are these?*

Asking and saying what things are; *this*, *that*, *these* and *those*

VOCABULARY AND SOUNDS

1  Listen and repeat.

pen books cassettes clock umbrella
watch wallet glasses bag keys

2 🔲 Listen and number.

1 - cassettes

3 Work in pairs. Look at the picture. Point
and check.

One   They're cassettes.

🔲 Now listen and check.

4 🔲 Read and listen.

/e/ p<u>e</u>n cass<u>e</u>tte Fr<u>e</u>nch t<u>e</u>n

/ɒ/ cl<u>o</u>ck w<u>a</u>tch w<u>a</u>llet d<u>o</u>ctor

/ɪ/ th<u>i</u>s <u>i</u>s s<u>i</u>x

Now say the words.

LISTENING AND SPEAKING

1 🔲 Listen and read. Point to
the objects.

A What's this?
It's a watch.

B What are these?
They're glasses.

C What's that?
It's an umbrella.

D What are those?
They're books.

2 Ask and say what things are in the
classroom.

22

FUNCTIONS

> **Asking and saying what things are**
>
> *What's this?*   *It's a watch.*
>
> *What's that?*   *It's an umbrella.*
>
> *What are these?*   *They're glasses.*
>
> *What are those?*   *They're books.*

1 Work in pairs.

Student A: Turn to Communication activity 4 on page 92.

Student B: Turn to Communication activity 17 on page 95.

2 Work in pairs. Ask and say what's in the picture below.

What's this?   *It's a football.*

What's that?   *It's a television.*

VOCABULARY AND SPEAKING

1 Translate. How many words are similar to your language?

> sandwich television pizza theatre
> football telephone coffee cinema tennis
> taxi bus burger video

2 Play *Word Bingo*.

> **Word Bingo**
>
> 1 Complete the chart with words from Lessons 1 – 10.
>
> 2 Work in groups of four or five. One of you says words from lessons 1 – 10.
>
> 3 Tick (✔) the words in your chart if they are there.
>
> 4 Are there five ticks (✔) in a line? Yes? Say *Bingo!*

telephone				
		group		
			football	
				American
	teacher			

Progress check 1–10

VOCABULARY

1 Put the words from Lessons 1 – 10 under the following headings: *nationalities, countries, jobs, numbers, family, classroom language.*

Nationalities: American

American ask correct sister doctor
English fifteen forty journalist
brother Japan listen one read
student teacher Thai Turkey
twenty-four

2 Write more words in the columns in 1.

3 Look at the pictures and complete the crossword with classroom words.

		1		c			
2				l			
		3		a			
			4	s	a	y	
			5	s			
		6		r			
		7	l	o	o	k	
		8	p	o	i	n	t
9				m			

4 Look at the words in Lessons 1 to 10 again. Choose words which are useful to you and write them in your *Wordbank* in the Practice Book.

GRAMMAR

1 Match the questions and answers.

1	What's your name?	a	No, they aren't.
2	How are you?	b	They're glasses.
3	Are they from London?	c	Sylvester Stallone.
4	What's this?	d	No, I'm not.
5	Who's your favourite actor?	e	My name's Steve.
6	Is she nineteen?	f	He's a teacher.
7	What are these?	g	Fine thanks, how are you?
8	How old are you?	h	Twenty-seven.
9	What's his job?	i	It's a book.
10	Are you married?	j	Yes, she is.

2 Punctuate.

1 whats her name
2 im very well
3 how do you spell book
4 is this a book
1 What's her name?

5 my names frank
6 where are you from
7 im from acapulco in mexico
8 no it isnt

3 Tick (✔) the correct sentence.

1 a He's English. ☐

   b His English. ☐

2 a His name's Frank. ☐

   b He's name's Frank. ☐

3 a What's you're name? ☐

   b What's your name? ☐

4 Are you married?

   a Yes, I am. ☐

   b Yes, I'm. ☐

5 a They're from Mexico. ☐

   b Their from Mexico. ☐

6 a We're English. ☐

   b Where English. ☐

4 Complete.

1 Is ____ your brother?   Yes, he ____.

2 What ____ those?   They ____ glasses.

3 Are you English?   Yes, we ____.

4 Are ____ French?   Yes, they ____.

SOUNDS

1 🔲 Listen and repeat.

/æ/ th<u>a</u>nk t<u>a</u>xi <u>a</u>ctor <u>A</u>merican

/e/ t<u>e</u>lephone p<u>e</u>n umbr<u>e</u>lla el<u>e</u>ven

/aɪ/ f<u>i</u>ve n<u>i</u>ne n<u>i</u>nety m<u>y</u> Un<u>i</u>ted

2 🔲 Listen and underline the stressed syllable.

Turkey Japan Thailand England seven eleven
thirty eighteen hundred cassette umbrella
telephone

3 🔲 Listen and repeat.

1 <u>What's</u> your <u>name</u>?   3 <u>How old</u> are <u>you</u>?

2 Are you from <u>London</u>? 4 <u>What's</u> your <u>job</u>?

READING AND LISTENING

1 Read the words of the song *Rock around the clock* by Bill Hailey and the Comets. Decide where these sentences go.

a eight, nine, ten, eleven, too

b twelve

c two, three and four,

d five and six and seven

e Five, six, seven o'clock, eight o'clock rock

One, two, three o'clock, four o'clock rock,
(1) ____
Nine, ten, eleven o'clock, twelve o'clock rock,
We're gonna rock around the clock tonight.
Put your glad rags on and join me hon'
We'll have some fun when the clock strikes one,
Chorus
We're gonna rock around the clock tonight
We're gonna rock, rock, rock til broad daylight
We're gonna rock, gonna rock around the clock tonight.
When the clock strikes (2) ____
If the band slows down, we'll yell for more
Chorus
When the chimes ring (3) ____
We'll be rockin' up in seventh heaven.
Chorus
When it's (4) ____
I'll be going strong and so will you,
Chorus
When the clock strikes (5) ____, we'll cool off, then
Start a rocking round the clock again.
Chorus

2 🔲 Listen and check.

25

11 | *How much are they?*

Talking about prices; position of adjectives; the definite article *the*

VOCABULARY

1 📼 Listen and repeat.

| jeans jacket shoes skirt shirt sweater |

2 Work in pairs. Look at the picture. Point and say.

3 Match the words in the box with the colours below.

| black blue white red green |

4 📼 Listen and match the clothes and the prices.

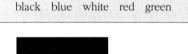
| pound pence |

Twelve pounds fifty pence = £12.50

£12.50
£40.99
£35.99
£27.99
£21.50
£50.00

5 Work in pairs and check your answers.

The jeans are twenty-seven pounds ninety-nine.

6 Add.

£4 + £5 =	£20 + £2.50 =
£12 + £6 =	£34.99 + £2 =
£15.50 + £2.50 =	£52.50 + £23.99 =

7 Work in pairs and check your answers.

LISTENING AND SOUNDS

1 📼 Listen and underline anything which is different.

CUSTOMER	How much are these red shoes?
ASSISTANT	They're £27.50.
CUSTOMER	And how much is that white sweater?
ASSISTANT	It's £15.
CUSTOMER	How much is this black jacket?
ASSISTANT	It's £50.
CUSTOMER	How much are these blue jeans?
ASSISTANT	They're £35.99.

2 Work in pairs and correct the conversation.
📼 Now listen again and check.

3 📼 Listen and repeat.
How much is this black jacket?
It's fifty pounds.
How much are these blue jeans?
They're thirty-five, ninety-nine.

FUNCTIONS AND GRAMMAR

> **Talking about prices**
> *How much is this black jacket?*
> *It's fifty pounds. (£50)*
> *How much are these blue jeans?*
> *They're thirty-five, ninety-nine. (£35.99)*
>
> **Position of adjectives**
> *the blue jacket*
> *the black jeans*
> *the white sweater*
> *NOT ~~the jacket blue~~*
> *NOT ~~the jeans black~~*
> *NOT ~~the sweater white~~*
>
> **The definite article** *the*
> *How much are the red shoes?*

1 Match the sentence with the picture.

a How much are they?
b How much is it?

1

2

2 Work in pairs. Act out the conversation in *Listening and sounds* activity 1.

3 Work in pairs.

Student A: Turn to Communication activity 5 on page 92.

Student B: Turn to Communication activity 18 on page 95.

4 Tick (✔) the correct sentence.

1 a How much are the shoes black?
  b How much are the black shoes?

2 a How much is the white shirt?
  b How much is the shirt white?

3 a How much is the green sweater?
  b How much are the green sweater?

LISTENING AND SPEAKING

1 Say what the objects in the photo are.

2 🔊 Listen and complete the chart.

a £30   b £0.35   c £3.50   d £2
e £10   f £40   g £5.99

	Britain	Your country
Big Mac		
Levi jeans		
Nike trainers		
Cola		

Now work in pairs and check your answers.

How much is a Big Mac in Britain?
It's ...

3 Complete the column with prices for your country. Work in pairs and ask and answer.

How much is a Big Mac in Thailand?
It's ....

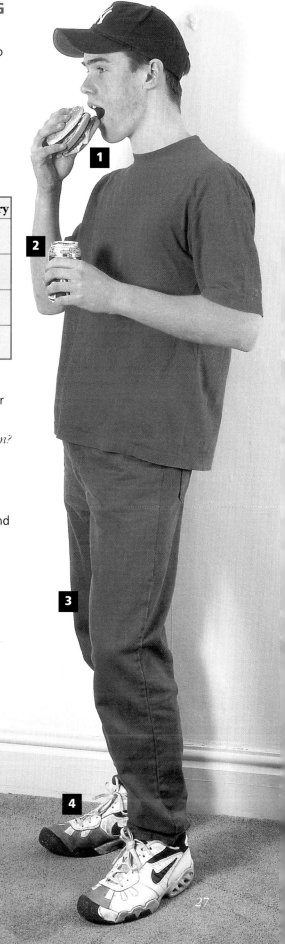

12 Where are Jane's keys?

Prepositions of place: *in, on, under*; possessive *'s*

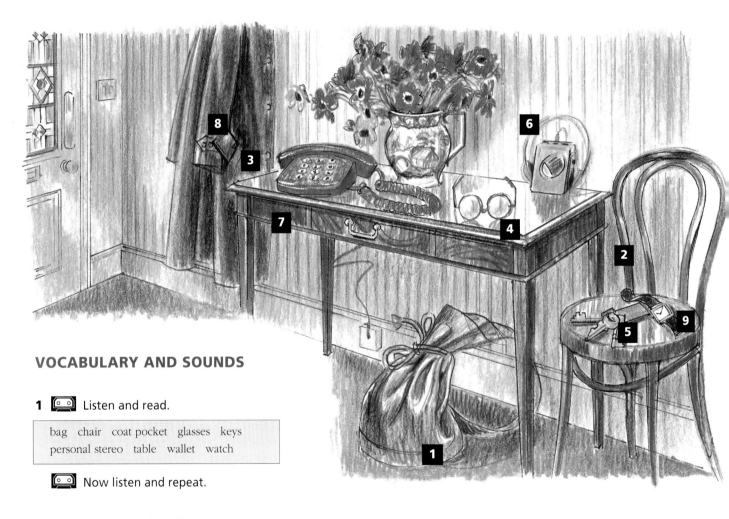

VOCABULARY AND SOUNDS

1 🔊 Listen and read.

> bag chair coat pocket glasses keys
> personal stereo table wallet watch

🔊 Now listen and repeat.

2 Work in pairs. Look at the picture.
Ask and say the words you know.

What's this?

It's a chair. And what are these?

They're keys.

LISTENING

1 🔊 Listen and match.

Jane	bag
Graham	glasses
Frank	keys
Joely	personal stereo
Nicola	watch
Tom	wallet

2 Match with the correct object in the picture.

a Jane's keys d Joely's bag

b Graham's wallet e Nicola's glasses

c Frank's watch f Tom's personal stereo

3 Match.

bag		
glasses	in	chair
keys	on	table
personal stereo	under	coat pocket
watch		
wallet		

[cassette] Now listen and check 2 and 3.

4 Tick (✔) the correct sentences.

1 Frank's watch is on the chair. ☐

2 Graham's wallet is on the table. ☐

3 Joely's bag is under the table. ☐

4 Tom's personal stereo is on the table. ☐

5 Nicola's glasses are under the chair. ☐

6 Jane's keys are on the chair. ☐

7 Nicola's glasses are in her coat pocket. ☐

8 Graham's wallet is in his coat pocket. ☐

9 Nicola's glasses are on the table. ☐

10 Frank's watch is under the table. ☐

GRAMMAR

Prepositions of place: *in, on, under*

*Jane's keys are **on** the chair.*

*Graham's wallet is **in** his coat pocket.*

*Joely's bag is **under** the table.*

Possessive 's

*Jane**'s** keys = the keys of Jane.*

*Graham**'s** wallet = the wallet of Graham.*

*Joely**'s** bag = the bag of Joely.*

---

1 Work in pairs and check your answers to *Listening* activity 4.

2 Look at *Listening* activity 4 and correct the wrong sentences.

2 Graham's wallet is in his coat pocket.

3 Complete.

1 **TOM** _____'s my personal stereo?

JOELY _____ on the table.

2 **JOELY** Where's _____ bag?

NICOLA It's under _____ table.

3 **GRAHAM** Where _____ my wallet?

JANE It's in _____ coat pocket.

[cassette] Now listen and check.

4 Look at 3 and write sentences.

Tom's personal stereo is on the table.

5 Tick (✔) the correct sentence.

1 a Anna's watch is in her coat pocket.

b Anna is watch is in her coat pocket.

2 a Where's Graham's wallet?

b Where's is Graham's wallet?

3 a Jane's keys on the table.

b Jane's keys are on the table.

4 a Nicola's glasses in her coat pocket.

b Nicola's glasses are in her coat pocket.

SPEAKING AND WRITING

1 Work in pairs.

Student A: Turn to Communication activity 7 on page 92.

Student B: Turn to Communication activity 19 on page 95.

Now write sentences describing the picture you completed.

The bag is under the table.

2 Choose three students and ask them to leave the classroom. Move six objects and ask the three students to say what is different.

Maria's bag is under the table.

13 | *We've got three children*

Have got; possessive adjectives: *our, your, their*

READING AND VOCABULARY

1 Read and match the texts with the photos.

A   Hello, my name's Maria and this is my husband Carlo. We've got three children. Our daughter's name is Laura and she's four. Our sons are Pablo and Octavio. Pablo is twelve and Octavio is ten.

B   Hello, I'm Carlos. I'm twelve. This is my family. I've got two sisters and one brother. My father's name is Javier. He's forty. My mother's name is Victoria. She's thirty-five. My sisters' names are Luisa and Teresa. They're eight and ten. My brother's name is Juan. He's five.

C   Hello, I'm Mehmet and this is my wife Sema. We've got two daughters, Leyla and Serap.

2 Complete.

> husband   wife   mother   father   son(s)
> daughter(s)   brother(s)   sister(s)   children

1 My _____ is Carlo.

2 Our _____ is Laura.

3 Our _____ are Pablo and Octavio.

4 Our _____ are Laura, Pablo and Octavio.

5 My _____ is Javier.

6 My _____ is Victoria.

7 My _____ are Luisa and Teresa.

8 My _____ is Juan.

9 My _____ is Sema.

10 My _____ are Leyla and Serap.

3 Complete the columns in the chart with names.

	husband	wife	father	mother	son (s)	daughter (s)	sister (s)	brother (s)
Maria								
Carlos								
Mehmet								

LISTENING AND SPEAKING

1 🔊 Listen and read.

INTERVIEWER	Are you married, Maria?
MARIA	Yes, I am. My husband's name is Carlo.
INTERVIEWER	Have you got any children?
MARIA	Yes, we have. We've got three children. Our daughter's name is Laura, and our sons' names are Pablo and Octavio.
INTERVIEWER	How old are they?
MARIA	They're twelve, ten, and four.

2 Decide where these sentences go.

a Paolo, Giovanni, and Patrizia.

b No, I haven't.

c Yes, I have. I've got two brothers and one sister.

d Yes I am.

Are you married, Marco?

(1) _____.

Have you got any children?

(2) _____.

Have you got any brothers or sisters?

(3) _____.

What are their names?

(4) _____.

🔊 Now listen and check.

GRAMMAR

> **Possessive adjectives**
> *Our* daughter is Laura.
> *Your* brothers are Pablo and Octavio.
> *Their* names are Laura, Pablo and Octavio.
>
> *Have got*
> Have you got any brothers and sisters?
> Yes, I have. I've got two brothers and one sister.
> Have you got any children?
> No, I haven't.

1 Answer the questions.

Is *our* the possessive adjective for *we, you* or *they*?

Is *their* the possessive adjective for *we, you* or *they*?

2 Complete.

I – *my*   he – ____   we – ____

you – ____   she – ____   they – ____

3 Tick (✔) the correct sentence.

1 a Are they're names Carla and Patrizia?
   b Are their names Carla and Patrizia?

2 a Our son is seven.
   b We're son is seven.

3 a Their son's name is Enrique.
   b They're sons name Enrique.

4 Complete.

1 Have you ____ any brothers or sisters?

2 Yes, I ____. I ____ got two brothers.

3 ____ you got any children?

4 No, I ____.

SOUNDS AND SPEAKING

1 🔊 Listen and repeat.

/ə/ husb*a*nd m*o*ther fath*er* daught*er*
broth*er* sist*er*

2 Complete the *You* line with names of members of your family.

	husband	wife	mother	father
You				
Your partner				
	son	daughter	brother	sister
You				
Your partner				

3 Work in pairs. Talk about your families. Ask and say what their names are. Complete the chart.

Have you got any brothers or sisters?

What are their names?

Are you married?

What's your wife's name?

14 | *She's got fair hair and blue eyes*

Talking about appearance and character; *has got*

READING AND VOCABULARY

1 📼 Listen and repeat.

hair eyes fair dark red blue brown black green

2 Write the words in the correct group.

hair – fair   red ...

eyes – blue ...

3 📼 Listen and repeat.

good-looking   friendly   pretty   nice quiet   tall   short

Now translate the words.

4 Put the words in 2 and 3 in two groups.

Appearance *– good-looking . . .*

Character *– friendly . . .*

5 Think about people you know and choose words to describe them. Now work in pairs and talk about these people.

quite   very

My brother's tall and quite good-looking. He's very friendly.

dark hair   green eyes

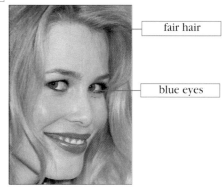

fair hair

blue eyes

LISTENING AND READING

1 📼 Listen and read.

JILL   Hey Sarah! We've got a new neighbour. His name's Mike.

SARAH   What's he like?

JILL   He's got dark hair and brown eyes. He's very good-looking and very friendly.

SARAH   Dark hair, brown eyes, good-looking and friendly! How old is he?

JILL   I don't know. Twenty, twenty-one.

SARAH   Has he got a brother?

JILL   I don't know.

2 Decide where these sentences go.

a She's got dark hair and green eyes.

b Yes, she has.

c What's she like?

d She's nice, but very quiet.

MIKE   Hey! Simon! We've got a very pretty neighbour.

SIMON   Really? What's her name?

MIKE   Jill.

SIMON   (1) _____

MIKE   (2) _____ She's very pretty.

SIMON   Green eyes! Has she got a friend?

MIKE   (3) _____

SIMON   What's she like?

MIKE   (4) _____

GRAMMAR AND FUNCTIONS

> **Talking about appearance and character.**
> *What's he like? What's she like?*
> *He's got dark hair and green eyes. He's very good looking.*
> *She's got fair hair and blue eyes. She's pretty.*
> *She's nice, but very quiet.*

1 Answer the questions.

Do you use *has got* with *we, you, they* or with *he, she, it?*

Do you use *have got* with *we, you, they* or with *he, she, it?*

2 Complete.

I	___		
we	___		
you	___		
they	___	got	fair hair.
he	___		
she	___		
it	___		

3 Work in pairs and check your answers to *Listening and reading* activity 2.

4 Write a description of someone you know. Say:

– what colour hair he/she's got

– what colour eyes he/she's got

– what he/she's like

Bruno has got dark hair and brown eyes. He's tall. He's quite quiet but nice.

5 Work in pairs.
Student A: Choose somebody in your class and describe them.
Student B: Say who it is.

She's got dark hair and brown eyes. She's got a red bag.

Is it Yıldız?

Yes!

READING AND SPEAKING

1 Read this part of an E-mail and complete the chart.

Dear Sue,
Hi, I'm Brad Summers from Chicago. I'm a doctor.
I'm 27 and I've got dark hair and blue eyes.
I've got two brothers, Joe and Bill, and a
sister, Judy. Joe is 25 and he's an engineer.
Bill's 23 and he's at college. Judy's 17 and
she's at school . . .

Name	Age	Appearance	Relation	Job
Brad	27			
Joe		"		
Bill		dark hair/ brown eyes	brother	student
Judy		"		

2 Work in pairs.

Student A: Turn to Communication activity 8 on page 93.

Student B: Turn to Communication activity 20 on page 96.

Sue		dark hair/ brown eyes		
Pippa	51	"		teacher
Henry	55		father	
James				
Sarah				

3 Work in pairs. Complete the chart for Sue.

WRITING

1 Copy and complete the chart above for you and your family.

2 Write a paragraph describing yourself and your family.

15 Stand up!

Imperatives

VOCABULARY AND LISTENING

1 🔲 Listen and repeat.

> window   door   book   light   cassette player
> pen   bag   coat

2 Work in pairs. Look at the picture. Point and say. Use all the words from 1.

3 Match the instructions with the people in the picture.

> Stand up.   Open your book.
> Turn the cassette player on   Pick your pen up.
> Come in.   Put your coat on.

1 – Stand up

4 Match the instruction with its opposite.

> Sit down.   Close your book.
> Turn the cassette player off.   Put your pen down.
> Go out.   Take your coat off.

Stand up – sit down

5 Make instructions with these verbs and the words in 1.

> open   close   turn on   turn off   pick up   put down

Open the window. Open the door.

6 Think about what you do when you start your English class. Number the instructions in 3 and 4.

1 – Come in. 2 – Take your coat off.

🔲 Now listen and check.

7 Match the sentences with the pictures.

a Don't talk!   b Don't look!   c Don't listen!

34

SOUNDS AND SPEAKING

1 📼 Listen and repeat.

/əʊ/   don't   close   open   coat

2 📼 Listen and read.

☐
Stand up.

☐
Sit down.

☐ ☐
Open your book.

☐ ☐
Close your book.

☐ ☐
Don't talk!

☐ ☐
Don't look!

☐ ☐
Pick your bag up.

☐ ☐
Put your pen down.

3 📼 Listen and follow the instructions.

4 Work in pairs. Give your partner instructions.

Close your book.

GRAMMAR

Imperatives		
Infinitive	**Imperative**	**Negative imperative**
look	*look*	*Don't look* *(= do not look)*
read	*read*	*Don't read*
write	*write*	*Don't write*
listen	*listen*	*Don't listen*

1 Look at the pictures below and write instructions.

2 Translate these words.

read   listen   write   underline   number
match   correct

READING AND WRITING

1 Read *The Students' Charter*. Which instructions are true for your English class?

The Students' Charter

☐ Don't speak English.
☐ Don't read in class.
☐ Put your bags under your chair.
☐ Do your homework.
☐ Say please and thank you.
☐ Don't talk to other students.
☐ Listen to your personal stereo.
☐ Stand up when your teacher speaks to you.
☐ Look out of the window.

2 Work in pairs and check your answers.

3 Correct the wrong instructions.

4 Write some more instructions for your classroom or your school.

Don't speak Turkish.

3.

5.

1.

2.

4.

6.

16 | *We live in a flat in Florence*

Present simple: regular verbs *I*, *we*, *you*, *they*; prepositions of place *in*, *to*

VOCABULARY AND SOUNDS

1 📻 Listen and repeat.

flat house office shop school

live work go

2 📻 Listen and repeat.

/æ/ flat match thank
/aʊ/ house how
/ɒ/ office shop clock
/uː/ school two student
/ɪ/ live tick six
/ɜː/ work turn Turkey

3 📻 Listen and repeat.

live in a flat
live in a house

work in an office
work in a shop

go to school
go to work

READING AND LISTENING

1 Match the sentences with the photos.

A I live in a flat in Florence.
B I work in an office in Tokyo.
C We go to school in Istanbul.

2 📻 Listen and read.

Hello, I'm Anna. I'm from Italy and I'm twenty-five. I'm a secretary and I work in an office. My husband's name is Bruno and we live in a flat in Florence. My mother's name is Francesca and my father's name is Alberto. My sister's name is Paola and she's sixteen. They live in a house in Fiesole.

3 Match the sentences with the photos in 1.

☐ Hello, I'm Kazuo.
☐ Hello, I'm Erol.
☐ I'm from Japan and I'm thirty-five.
☐ I'm from Turkey and I'm sixteen.
☐ I'm a student.
☐ I'm a journalist and I work in an office in Tokyo.
☐ My sister's name is Belma and we go to school in Istanbul.
☐ My wife's name is Michiko and we live in a flat in Ichikawa.
☐ We live with our parents in a flat in Galata.
☐ We've got two children.
☐ My father is an engineer and my mother is a secretary.
☐ Our son's name is Koji and our daughter's name is Miki.
☐ They work in an office in Beyoglu.
☐ They go to school in Funabashi.

📻 Now listen and check your answers.

GRAMMAR

> **Present simple: regular verbs *I, we, you, they***
>
> *I*
> *We*   *live*
> *You*
> *They*
>
> **Prepositions of place**
>
> *In*   *I work **in** an office **in** Tokyo. I live **in** a flat **in** Florence.*
> *To*   *I go **to** school.*

1 Complete.

1 I _____ in a flat in Paris.

2 I _____ in an office in London.

3 We _____ to school in New York.

4 They _____ in a flat in Paris.

5 We _____ in an office in London.

6 I _____ to school in New York.

7 They _____ in an office in London.

8 They _____ to school in New York.

9 We _____ in a flat in Paris.

2 Put the words in the right order and make sentences.

1 school Paris in I to go

2 school we go to New York in

3 live flat we in in a Venice

4 work I office Tokyo in an in

5 we live in a in Rio house

6 I in an office work

1 I go to school in Paris.

3 Complete.

1 We live in a house _____ Buenos Aires.

2 I work _____ a shop in Istanbul.

3 I go _____ school _____ Lyon.

4 They work _____ an office _____ Seoul.

5 We work _____ a shop _____ London.

6 They go _____ school _____ Athens.

READING AND WRITING

1 Read and answer *What's my name?*

> ### What's my name?
>
> 1 I'm from the USA.
> 2 I live in Washington.
> 3 I live with my wife and daughter.
> 4 I work in the USA and around the world.
> 5 We live in the White House.
> 6 I'm President of the USA.
> 7 My name's _____.

2 Write sentences about someone famous. Use *I* and *we*.

I live in Buckingham Palace.

I live with my husband and my dogs.

Now work in pairs and play *What's my name*.

3 Write a paragraph about where you live.

Hello, I'm Marcella. I'm from Argentina and I'm twenty. I live in Rosario with my father and mother …

17 *What's the time?*

Telling the time (1); present simple: *have*; prepositions of time: *at, in*

SPEAKING AND VOCABULARY

1 📼 Listen and repeat.

> o'clock

What's the time?

A It's eight o'clock.

B It's eleven o'clock.

C It's ten o'clock.

2 Match the sentences in 1 with the correct time.

3 Work in pairs. Ask and answer about the times.

What's the time? It's 10.00.

4 📼 Listen and tick (✔).

1   a 3 o'clock ☐   b 5 o'clock ☐
2   a 7 o'clock ☐   b 11 o'clock ☐
3   a 6 o'clock ☐   b 9 o'clock ☐
4   a 12 o'clock ☐   b 1 o'clock ☐

5 📼 Listen and repeat.

> have breakfast   have lunch   have dinner

> morning   afternoon   evening

6 Complete sentences about you.

1   I have breakfast at ____ in the ____.
2   I have lunch at ____ in the ____.
3   I have ____ at ____ in the evening.

7 Look and read.

> am   pm

Seven o'clock in the morning = 7am
One o'clock in the afternoon = 1pm
Seven o'clock in the evening = 7pm

Now rewrite these times with *am* or *pm*.

1   Four o'clock in the morning.
2   Two o'clock in the morning.
3   Eleven o'clock in the evening.
4   Three o'clock in the afternoon.

LISTENING AND READING

1 📼 Listen and read. Say which country you see in the photo.

'In Britain, we have breakfast at about 8 am. Lunch is at 1pm and dinner is at 6 pm.'

'In Spain, we have breakfast at 7am and lunch at 2 or 3 pm. Dinner is at 10 pm.'

'In Thailand we have breakfast at 6 am and lunch at 11 am. Dinner is at about 6 pm.'

2 Work in pairs and check your answer to 1.

3 Work in groups of three.

Student A: Turn to Communication 9 on page 93.

Student B: Turn to Communication 21 on page 96.

Student C: Turn to Communication 27 on page 97.

4 Work together and complete the chart.

speaker	country	breakfast	lunch	dinner
1				
2				
3				

FUNCTIONS AND GRAMMAR

> **Telling the time (1)**
> *What's the time?   It's one o'clock.*
>
> **Prepositions of time**
> **At**   *at seven o'clock*   *at twelve o'clock*   *at five o'clock*
> **In**   *in the morning*   *in the afternoon*   *in the evening*
>
> **Present simple: *have***
> *I*
> *We*   *have lunch at two o'clock in the afternoon.*
> *You*
> *They*

1 Write.

1 In Britain they have breakfast at eight o'clock in the morning. They have lunch at _____.

2 In Spain they have breakfast _____.

3 In Thailand they _____.

4 In my country, we _____.

2 Tick (✔) the correct sentence.

1 a What's the time?   b What the time?

2 a Seven o'clock.   b It seven o'clock.

3 a I'm breakfast at eight o'clock.
   b I have breakfast at eight o'clock.

4 a We have lunch at one o'clock.
   b We have lunch one o'clock.

3 Complete.

1 We have breakfast ____ seven o'clock ____ the morning.

2 I have lunch ____ one o'clock ____ the afternoon.

3 They have dinner ____ nine o'clock ____ the evening.

SPEAKING

1 Work in pairs. Look at the map below. It's twelve o'clock in London. Ask and say.

What's the time in New York?

It's seven o'clock in the morning. What's the time in Hong Kong?

It's eight o'clock in the evening.

2 Find out what time people in your class have breakfast, lunch and dinner.

Name	Nationality	breakfast	lunch	dinner
Marco	Italian	8 am	1-3 pm	8 pm

In Italy, they have breakfast at 8 ...

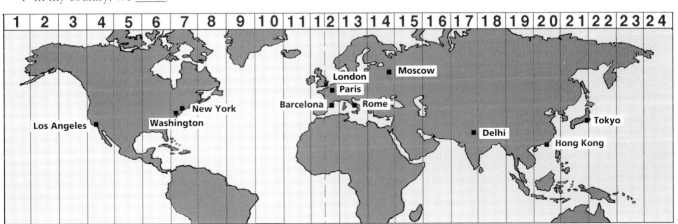

18 | *I don't like Monday mornings*

Present simple: negatives; preposition of time: *on*

VOCABULARY AND SOUNDS

1 🔊 Listen and repeat.

Monday	Tuesday	Wednesday
Thursday	Friday	Saturday	Sunday

2 Complete.

1 Today is ____.

2 Tomorrow is ____.

3 🔊 Listen and repeat.

on Monday
on Tuesday
on Wednesday
on Thursday

on Monday morning
on Tuesday afternoon
on Wednesday evening

in the morning
in the afternoon
in the evening

4 Match the verbs with the photos.

watch television   go to the cinema
see friends   go shopping   write letters
go for a walk

5 Match.

listen to   a newspaper
read   a letter
write   homework
correct   music
                   a book

listen to music, ...

LISTENING AND READING

1 🔊 Listen and read.

Hello! My name's Fiona and I'm a teacher. I work in a school in the mornings and in the afternoons from Monday to Friday. In the evenings I watch television, listen to music, read, or write letters. In Britain we don't work at the weekend, so on Saturdays I go shopping in the morning or in the afternoon, and, in the evening, I go to the cinema or see friends.

I don't live with my family so on Sunday morning I go to their house. We have Sunday lunch and then my father and brother watch the football on television. My sister and I don't like football so we go for a walk with Mum on Sunday afternoon. On Sunday evening, I correct my students' homework. I go to bed at 9 and get up at 7.30 am. I don't like Monday mornings!

2 Write what Fiona does.

	Morning	Afternoon	Evening
Monday ➡ Friday			
Saturday			
Sunday			

GRAMMAR

Present simple: negatives

I
We *don't (= do not) like Monday mornings.*
You
They

Preposition of time

on *Monday* **on** *Tuesday* **on** *Wednesday*

on *Monday morning* **on** *Tuesday afternoon*

on *Wednesday evening*

1 Complete.

1 We're students. We ____ work in an office.

2 I live in London. I don't ____ in Paris.

3 They ____ have breakfast at seven o'clock. They have breakfast at eight.

4 We ____ watch television on Saturday. We go to the cinema.

5 I ___ shopping on Saturday. I don't go shopping on Sunday.

6 I don't like tennis. I ____ football.

2 Complete.

1 I don't work ____ Saturday and Sunday.

2 ____ Sunday morning, I read the newspaper.

3 We don't watch television ____ the morning.

4 ____ the evening I write letters.

5 ____ Saturday afternoon they play football.

6 We see friends ____ Saturday.

3 Tick (✔) the sentences which are true for you. Correct the false ones.

1 I like Monday mornings.

2 We work on Sunday.

3 My parents live in France.

4 I go to the cinema on Monday morning.

5 We go shopping on Thursday evening.

6 My friends play football on Wednesday.

1 I don't like Monday mornings.

WRITING AND SPEAKING

1 Copy the chart in *Listening and reading* activity 2 and complete it for you.

2 Work in pairs.

Student A: Say a day.

Student B: Say what you do.

3 Write a paragraph about what you do during the week.

I go to school/work from Monday to Friday. On Monday evening, I ...

19 *Do you like running?*

Yes/no questions and short answers

VOCABULARY AND SOUNDS

1 🔊 Listen and repeat.

> football   tennis   volleyball   table tennis   skiing
> running   basketball   gymnastics   swimming
> baseball   sailing

2 Say the words you know.

3 Work in pairs.

Student A: Turn to Communication activity 11 on page 93.

Student B: Turn to Communication activity 23 on page 96.

4 Complete.

> team sport   individual sport

> *team sport*   football ...
> *individual sport*   tennis ...

5 Write the words in 1 under these headings: *I like* and *I don't like*.

> like

I like: football, tennis...

I don't like: badminton, skiing...

LISTENING AND WRITING

1 🔊 Listen and read.

DAVE   Do you like swimming?

JACK   Yes, I do. I like swimming very much.

DAVE   Do you like running?

JACK   No, I don't.

2 Complete.

TIM   Do you _____ gymnastics, Gwen?

GWEN   Yes, I _____. I like gymnastics very much.

TIM   _____ you like basketball?

GWEN   _____, I don't.

3 Number the sentences in the right order.

☐ Alison   No, I don't.

☐ James   Do you like skiing?

☐ James   Do you like volleyball, Alison?

☐ Alison   Yes, I do. I like volleyball very much.

🔊 Now listen and check 2 and 3.

4 Tick (✔) the sports Jack, Gwen and Alison like. Put a cross (✘) by the sports they don't like.

	Jack	Gwen	Alison
football			
tennis			
volleyball			
table tennis			
skiing			
running			
basketball			
gymnastics			
swimming			

GRAMMAR

Yes/no questions	Short answers
Do you like swimming?	*Yes, I do.*
Do you like table tennis?	*No, I don't. (= do not)*
Do you live in Berlin?	*Yes, I do. No, I don't.*
	Yes, we do. No, we don't.
Do they work in London?	*Yes, they do. No, they don't.*

1 Tick (✔) the correct answer.

1 Do you like tennis?
  a Yes, I do.   b Yes, I like.

2 Do you like volleyball?
  a No, I no like.   b No, I don't.

3 Are you married?
  a No, I am not.   b No, I don't.

4 Are they from London?
  a Yes, they are.   b Yes, they do.

5 Do they work in Paris?
  a Yes, they do.   b Yes, they work.

6 Do you live in Berlin?
  a Yes, we do.   b No, they don't.

2 Write five questions.

			in a flat?
		live	tennis?
Do	you	work	in a house?
		like	in an office?
			football?

3 Work in pairs. Ask and answer your questions from 2.

4 Answer the questions.

1 Do you like swimming?

2 Do you live in London?

3 Do you have breakfast at five o'clock in the morning?

4 Do you work in an office?

5 Are you married?

6 Are you a doctor?

7 Are you French?

8 Are you nineteen?

5 Work in pairs. Act out the conversations in *Listening and writing activities* 2 and 3.

SPEAKING AND WRITING

1 Find:

Four people who like swimming. Write their names here.

_____

_____

Four people who like basketball. Write their names here.

_____

_____

Four people who like gymnastics. Write their names here.

_____

_____

Four people who like football. Write their names here.

_____

_____

2 Find someone who likes the same sports as you.

Do you like swimming? Yes, I do.

3 Write a poster about sports.

Say what sports you like.
We like basketball.

Say what your favourite team OR player is.
Our favourite team is …

Say what the favourite sport in your country is.
People in my country like …

20 | *She likes her job*

Telling the time (2); present simple: *he, she, it*

VOCABULARY AND SPEAKING

1 🔊 Listen and repeat.

a quarter past   a quarter to   half past

What's the time?
A A quarter past one.
B Half past two.
C A quarter to three.

2 Match the sentences in 1 with the correct time.

3 Work in pairs. Ask and answer about the other times.

What's the time? It's a quarter past four.

4 🔊 Listen and tick (✔).

1	a 4.30 ☐	b 5.30 ☐	
2	a 6.45 ☐	b 6.30 ☐	
3	a 12.15 ☐	b 12.45 ☐	
4	a 5.30 ☐	b 5.45 ☐	
5	a 7.30 ☐	b 7.45 ☐	
6	a 11.15 ☐	b 11.30 ☐	

5 🔊 Listen and repeat.

leave   arrive   go home   start
visit   finish

6 Match the sentences with the pictures.

a Joe arrives at work at nine o'clock.

b He starts work at a quarter past nine.

c He goes home at a quarter to nine.

d He visits friends at six o'clock on Fridays.

e He finishes work at a quarter past five.

f He leaves home at half past eight.

7 Complete sentences about you.

1 I leave home at ____

2 I arrive at work/school at ___

3 I start work at ___

4 I finish work at ____

5 I go home at ____

6 I visit friends at/on ____

8 Work in pairs. Say your answers to 7.

I leave home at eight o'clock.

READING

Read the passage and complete the sentences.

Esther lives in a flat in Boston with her husband and their two sons. She works in a shop from Monday to Saturday. She has breakfast at 6.30 in the morning. She leaves at 7.30, and arrives at work at 8 am.

She has lunch at 12.30 in the afternoon, and starts work at 1.30 pm. She finishes work at 6 in the evening and goes home. She works on Saturday, but she likes her job.

Her husband works in a school in Boston. On Saturday afternoon, he goes shopping with their sons and on Sunday they visit Esther's parents. They have a house in Cambridge.

1 Esther _____ in a flat in Boston.

2 She _____ in a shop.

3 She _____ breakfast at 6.30.

4 She _____ at 7.30.

5 She _____ at work at 8 am.

6 She _____ lunch at 12.30 pm.

7 She _____ work at 6 pm.

8 Her husband is a teacher. He _____ in a school in Boston.

9 He _____ shopping with their sons on Saturday afternoon.

10 They _____ Esther's parents on Sunday.

GRAMMAR

> **Telling the time (2)**
> *What's the time? It's a quarter past one.*
> *It's half past two.*
> *It's a quarter to three.*
>
> **Present simple: *he, she, it***
> *Esther lives in a flat in Boston.*
> *Her husband works in a school.*
> *The train leaves at 7.30 am.*
> *It arrives at 8 am.*

1 Write the third person.

1 leave arrive work live start visit go finish have
leaves,...

2 Tick (✔) the correct sentence.

1 a She works in a shop.   b She work in a shop.
2 a He's live in Boston.   b He lives in Boston.
3 a Its leaves at 7.30.   b It leaves at 7.30.
4 a He go home at 6 pm.   b He goes home at 6 pm.
5 a Her parents live in Cambridge. b Her parents lives in Cambridge.
6 a She have two sons.   b She has two sons.

LISTENING AND WRITING

1 📼 Listen and tick (✔).

	Sarah		**Mark**	
live	☐ Spain	☐ Italy	☐ Mexico City	☐ Acapulco
work	☐ office	☐ school	☐ shop	☐ hospital
have breakfast	☐ 7.30 am	☐ 8 am	☐ 6.30 am	☐ 7.45 am
start work	☐ 3 pm	☐ 2 pm	☐ 6.45 am	☐ 8 am
finish work	☐ 9 pm	☐ 11 pm	☐ 6 pm	☐ 7pm
go shopping	☐ mornings/ Saturdays	☐ mornings/ Sundays	☐ Fridays	☐ Saturdays
visit parents/friends	☐ Saturdays/ Sundays	☐ Sundays/ Mondays	☐ Saturdays	☐ Sundays

2 Work in pairs. Check 1.

3 Write a paragraph about Sarah and Mark.

Sarah lives in Italy.

SPEAKING

1 Work in pairs. Turn to Communication activity 36 on page 99.

2 Work with another pair. Check your information.

Her name is Mary Ward.

Progress check 11–20

VOCABULARY

1 Tick (✔) the correct answer.

1 A *sweater* is
  a a place ☐
  b a member of the family ☐
  c an item of clothing ☐

2 A *wallet* is
  a a personal possession ☐
  b food ☐
  c a day of the week ☐

3 A *mother* is
  a a drink ☐
  b an eye colour ☐
  c a member of the family ☐

4 *Fair* is
  a a hair colour ☐
  b an item of clothing ☐
  c food ☐

5 *Good looking* is
  a a word to describe character ☐
  b a word to describe appearance ☐
  c a personal possession ☐

6 *Stand up* is
  a an instruction ☐
  b a drink ☐
  c a personal possession ☐

7 A *flat* is
  a a place where people live ☐
  b a day of the week ☐
  c a word to describe character ☐

8 *Thursday* is
  a a day of the week ☐
  b an item of clothing ☐
  c a personal possession ☐

9 *Lunch* is
  a a place where people live ☐
  b food ☐
  c a day of the week ☐

10 *Running* is
  a a day of the week ☐
  b a sport ☐
  c a drink ☐

2 Think of other words for all the categories in 1.

3 Look at the words in Lessons 11 to 20 again. Choose words which are useful to you and write them in your *Wordbank* in the Practice Book.

GRAMMAR

1 Write questions and answers.

1 wallet – £50.00

2 jacket – £39.00

3 jeans – £12.99

4 sweater – £25.50

5 skirt – £20.00

6 shirt – £25.00

1 How much is that wallet?

It's fifty pounds.

2 Look at the picture and say where the things are.

The telephone is on the table.

3 Complete the passage about the Smiths.

Name	George	Judy	Terri
Family	son	daughter	daughter
Appearance	fair hair,	dark hair,	dark hair,
	good looking	brown eyes	green eyes

The Smiths ____ got three ____. Their ____ name is George. He's got ___ hair. and he's very ____ ____. ____ daughters' ____ are Judy and ____. Judy has ____ hair and ____ eyes. Terri has ____ hair and ____ eyes.

4 Match the instructions with their opposites.

1 Come in.	a Close your book.
2 Stand up. '	b Go out.
3 Open your book.	c Sit down.
4 Take your coat off.	d Put your coat on.

5 Say what the time is.

6 Complete the sentences so they're true for you.

1 I live in a ____ in ____.
2 I work/go to school in ____.
3 My parents live in ____.
4 We have breakfast at ____.
5 We have dinner at ____.
6 I start school/work at ____.
7 I finish school/work at ____.
8 I like ____.

7 Work in pairs. Look at your partner's answers to 7. Rewrite them like this.

Pierre lives in a flat in Paris. He goes to school in ....

SOUNDS

1 🔊 Listen and repeat.

/ɑː/ dark glasses past
/ʌ/ son mother brother
/ɜː/ skirt shirt work Thursday
/eə/ hair fair their

2 🔊 Listen and underline the stressed syllable.

sweater jacket glasses father wallet

Now say the words aloud.

3 🔊 Listen and underline the stressed words.

1 I've got a brother and two sisters.
2 He's got fair hair and blue eyes.
3 I live in a house in London.
4 It's half past eight.
5 I don't like Monday mornings.
6 He likes his job.

Now say the sentences aloud.

LISTENING

1 🔊 You're going to hear *Three Little Birds*, by Bob Marley. The lines are in the wrong order. Listen and number the lines in the order you hear them.

☐ Rise up this morning
☐ Don't worry about a thing
☐ Smiled with the rising sun
☐ Singing 'This is my message to you.'
☐ 'Cause every little thing's gonna be all right
☐ Singing 'Don't worry about a thing
☐ Singing sweet songs of melodies pure and true
☐ 'Cause every little thing's gonna be all right'
☐ Three little birds beside my doorstep

2 Work in pairs and check your answers.

🔊 Now listen again and check.

21 *Does she go to work by boat?*

He/she/it; Yes/no **questions and short answers;** *by*

VOCABULARY AND SOUNDS

1 📻 Listen and repeat.

> train car bus taxi boat bicycle

> walk

> Asia Europe America

2 Turn to Communication activity 12 on page 94 and check you know what the words in 1 are.

3 Work in pairs. Say what means of transport you can see in the photo.

4 📻 Listen and repeat.

/eɪ/ train play eight say
/ɑː/ car ask father
/ʌ/ bus number mother
/æ/ taxi bag flat have
/əʊ/ boat go don't
/aɪ/ bicycle time arrive
/ɔː/ walk four quarter

5 📻 Listen and repeat.

car go to work
Do you go to work by car?

bus go to work
Do you go to work by bus?

walk work
Do you walk to work?

READING AND LISTENING

1 Read and complete the chart for Mehmet.

SALLY   Good morning! This is Reward Radio. My name's Sally Finch and today on *Going to work ... going home* we're in Turkey on the Galata Bridge in Istanbul. Turkey is in Europe and in Asia. The Galata Bridge is in Europe, but Asia is over there. It's eight o'clock on Monday morning and this is Mehmet. Mehmet, do you live here?

MEHMET   No, I don't. I live in Asia.

SALLY   Do you work here?

MEHMET   Yes, I do.

SALLY   So you live on the Asian side but you work in Europe?

MEHMET   Yes.

SALLY   Do you go to work by car?

MEHMET   No. I go to work by boat.

SALLY   And are you married?

MEHMET   Yes. My wife's name is Belma.

SALLY   Does Belma work in Istanbul?

MEHMET   Yes, she does.

SALLY   And does she go to work by boat?

MEHMET   No, she doesn't. She goes to work by car.

SALLY   Thank you Mehmet. ... Excuse me!

2 Complete.

1 ____ Mehmet ____ in Asia?

2 ____ Belma ____ to work by boat?

3 ____ Leyla ____ in a shop?

4 ____ Mustafa ____ lunch at home?

5 ____ John ____ his parents on Sunday?

6 ____ Mary ____ shopping on Saturday?

3 Answer the questions in 2.

1 Yes   2 No   3 Yes   4 No   5 Yes   6 No

Yes, he does.

SPEAKING

1 Work in pairs and act out the conversation in *Reading and listening* activity 1.

2 Act out the conversation again, but talk about your journey to school/work.

3 Work with other students. Find someone who:

– goes to work/school by car

– goes to work/school by taxi

– goes to work/school by train

– goes to work/school by bicycle

– goes to work/school by bus

– walks to school/work

Now write their names.

4 Work in pairs.

Student A: Look at Student B's list of names. Ask how they go to school/work. You score one point for each *Yes* answer.

Does Piotr go to school by taxi?

Student B: Show your list of names to Student A. Answer the questions.

Yes, he does.

	goes to work by
Mehmet	
Belma	
Leyla	
Mustafa	
John	
Mary	

2 🔲 Listen to the rest of the programme and complete the chart.

GRAMMAR

He/she/it; Yes/no questions

Does Mehmet go to work by boat?	*Yes, he does.*
Does Belma go to work by bus?	*No, she doesn't.*
Does he go to work by car?	*Yes, he does.*
Does she walk to work?	*No, she doesn't.*

By

by car by train by bus by taxi

---

1 Work in pairs and check your answers to *Reading and listening* activity 1.

Does Mehmet go to work by boat? Yes, he does.

22 *What do they eat in Morocco?*

Present simple: *Wh-* questions

VOCABULARY AND SOUNDS

1 Listen and repeat.

milk yoghurt rice beef
chicken lamb orange lemon
apple wine water juice potato
tomato bread tea coffee beer

2 Work in pairs.

Student A: Turn to Communication activity 13 on page 94.

Student B: Turn to Communication activity 24 on page 96.

3 Ask and say what the food and drink is.

What's this?

Milk. And what's this?

4 Complete.

food drink

Food	Drink
orange ...	water ...

5 Put the food under these headings.

vegetable meat fruit

Which words don't go under these headings?

READING AND LISTENING

1 Read and match the questions and answers.

Food and drink around the world.

1 What food do you eat in your country?

2 What do you drink?

3 When do you have the main meal of the day?

4 What do you have for breakfast?

A The main meal of the day is lunch. We have lunch at one o'clock in the afternoon. During the week, if you work, the main meal is dinner.

B In Argentina we eat meat, especially beef.

C We have coffee and bread.

D We drink beer and wine with our meals, or water and juice.

 Now listen and check.

2 Complete the *Argentina* column with the letter corresponding to the correct answer.

Question	Argentina	Morocco	India
1	*B*		
2			
3			
4			

GRAMMAR

> *Wh-* questions
> *What food do you eat in your country?*
> *When do you have the main meal of the day?*

1 Work in groups of three.

Student A: Turn to Communication activity 14 on page 94.

Student B: Turn to Communication activity 25 on page 97.

Student C: Turn to Communication activity 29 on page 97.

2 Work together and complete the Morocco and India columns of the chart with the correct information.

3 Tick (✔) the correct sentence.

  1 a When do you drink tea?
    b When you drink tea?

  2 a What's he have for breakfast?
    b What does he have for breakfast?

  3 a When does she eat?
    b When's she eat?

  4 a What do you have for breakfast?
    b What you have for breakfast?

4 Write.

  1 what/he/have for breakfast?

  2 when/she/have lunch?

  3 what/you/eat?

  4 what/she/drink?

  5 when/they/have the main meal?

  6 what food/eat in your country?

  What does he have for breakfast?

SPEAKING AND WRITING

1 Work in pairs. Talk about food and drink in your country. Ask and answer the questions in *Listening* activity 1.

2 Write a short paragraph describing food and drink in your country.

  In Italy, we eat pasta and pizza.

3 Complete.

Find someone who:	Name
drinks tea in the morning	_____
likes beer	_____
doesn't like meat	_____
has a sandwich for lunch	_____
eats pasta	_____
doesn't drink tea	_____
likes French fries	_____
eats burgers for breakfast!	_____

  What do you drink in the morning? Tea.

4 Work in groups. Ask and say.

  What does Marco drink in the morning?
  He drinks tea.

23 *I don't like lying on the beach*

Like + -ing; present simple: negatives

Lying on the beach   Eating in restaurants   Skiing   Sightseeing   Writing postcards

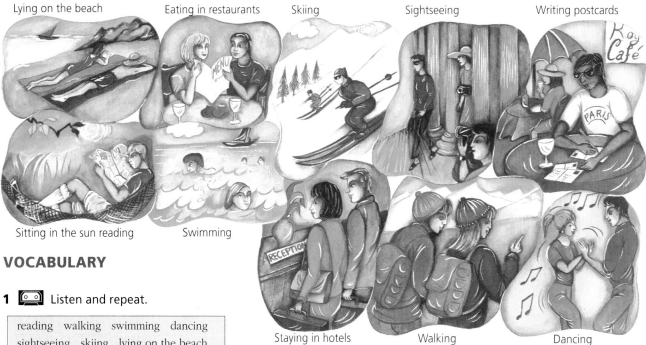

Sitting in the sun reading   Swimming

Staying in hotels   Walking   Dancing

VOCABULARY

1 📼 Listen and repeat.

> reading walking swimming dancing
> sightseeing skiing lying on the beach
> eating in restaurants staying in hotels
> writing postcards sitting in the sun

2 Work in pairs. Look at the pictures. Point and say.

3 Write the words in two columns: *I like* and *I don't like*.

I like: reading, swimming...

I don't like: walking...

READING AND LISTENING

1 Read the passages and say who the people in the photo are.

Going on Holiday

Where do you like going on holiday?
What do you like doing on holiday?
What don't you like doing on holiday?
We talk to people about going on holiday.

'I like walking and sightseeing. I don't like lying on the beach. My girlfriend likes sightseeing, but she doesn't like walking.' *Mick.*

'I don't like going on holiday with my family, so I usually go away with my friends, to Spain or to Turkey. We have a great time.' *Dave*

'We don't like staying in hotels. We have friends in the Rocky Mountains in the USA so we go skiing with them.' *Carrie*

'I don't like writing postcards but my parents like to get them, so I write a card to them, and a card to friends.' *Colin*

'I go to Europe during my college holidays. I don't like staying in America. I like travelling by train. I also like walking and sightseeing.' *Brad*

'We go to the sea. I don't like swimming but I like sitting in the sun and reading.' *Patrizia*

2 Say if these statements are true or false.

1 Mick likes lying on the beach.

2 Dave likes going on holiday with his family.

3 Carrie and her husband like staying in hotels.

4 Colin doesn't like writing postcards.

5 Brad likes going to Europe.

6 Patrizia likes swimming.

3 🔲 Listen to Gary and Margaret talking about what they like and don't like doing on holiday. Put a tick (✔) by the things they like doing and a cross (✗) by the things they don't like doing.

	Gary	Margaret
reading		
walking		
swimming		
dancing		
sightseeing		
skiing		
lying on the beach		
eating in restaurants		
staying in hotels		
writing postcards		
sitting in the sun		

4 Work in pairs and check your answers.

Gary likes sightseeing but Margaret doesn't.

🔲 Now listen again and check.

GRAMMAR

Like + -ing
I like swimming.
Patrizia likes sitting in the sun.
Mick likes walking and sightseeing.

Present simple: negatives
I don't like lying on the beach.
She doesn't like walking.
We don't like staying in hotels.

1 Work in pairs. Check your answers to *Vocabulary* activity 3.

I like swimming but I don't like sightseeing.

2 Correct the false statements in *Reading and listening* activity 2.

Mick doesn't like lying on the beach.

3 *Doesn't* or *don't?* Choose the correct sentences.

1 I doesn't/don't like swimming.

2 He doesn't/don't like staying in hotels.

3 We doesn't/don't like sightseeing.

4 They doesn't/don't like reading.

5 She doesn't/don't like dancing.

6 You doesn't/don't like walking.

SPEAKING AND WRITING

1 Work in pairs. Ask and answer the questions in *Reading and listening* activity 1.

2 Write a paragraph about your partner's holiday likes and dislikes. Read it to the class.

Marco likes going on holiday to Europe. He likes …

There is/are; any

VOCABULARY AND SOUNDS

1 Listen and repeat.

> kitchen   bathroom   bedroom   living room
> garden   dining room   hall

> downstairs   upstairs   front   back

> small   large

2 Read and label the rooms.

The *kitchen, living room* and *dining room* are downstairs.

The *dining room* is at the front.

There is a door from the hall to the *kitchen*.

The *living room* has a window. You see the garden.

There are two *bedrooms upstairs*.

The *bathroom* is upstairs, at the back.

3 Work in pairs. Ask and answer.

What rooms do you have in your home?

4 Listen and repeat.

> cooker   cupboard   armchair   sofa   bed   shower

5 Work in pairs. Look at the picture. Point and say these words.

6 Work in pairs. Say what's in each room.

kitchen – cooker, table ...

 Now listen and check.

54

READING AND LISTENING

1 📼 Listen and read. Say how many rooms there are.

'Hello and welcome to *Through the keyhole*, the game where I describe a house and you decide who lives here. Today I'm in a very large and beautiful house in the country. Downstairs there's a living room, a dining room with a window onto a large garden, and a kitchen. In the dining room there's a table and chairs for four people. Upstairs there are five bedrooms and three bathrooms. In the kitchen there's a cooker, a table, and five chairs. There's lots of food – fruit, vegetables but no meat. In the bedroom at the back, there's a large bed, and a television. There aren't any chairs. There's a cupboard with jeans, jackets, skirts and shoes. There's also a tennis racquet. At the moment, we're downstairs in the living room, there's a radio and there are some books, but there isn't a television. There are three sofas and four armchairs, so he or she likes having friends here. There's a table here with lots of photos of the man or woman who lives here, with his or her family.'

2 📼 Listen to descriptions of three people. Who lives in the house?

George Mandelson ☐   Angie Ashton ☐   Frances Peters ☐

3 Work in pairs. Check your answer to 2.

GRAMMAR

There is/are

There's a garden. There are five bedrooms.
Are there any chairs in the dining room?
Yes, there are.   No, there aren't. (= there are not)
Is there a shower?
Yes, there is.   No, there isn't. (= there is not)

Any

| *Plural negative* | *There aren't any cupboards in the bathroom.* |
| *Plural questions* | *Are there any chairs in the kitchen?* |

1 Tick (✔) the correct sentence.

1 a Is there a garden?
  b Are there a garden?

2 a There are two chairs.
  b There is two chairs.

3 a There isn't any plants.
  b There aren't any plants.

4 Is there a shower?
  a Yes, there are.
  b No, there isn't.

5 Are there any cupboards?
  a Yes, there are.
  b No, there isn't.

6 Is there a television?
  a Yes, there's.
  b No, there isn't.

2 Look at the picture and write sentences.

1 In the living room, there's ...

2 In the kitchen, there's a table ...

3 In the bedroom, ...

4 In the bathroom, ...

3 Work in pairs. Ask and answer questions about the house in the picture.

Are there any chairs in the dining room?
Yes, there are.

SPEAKING AND WRITING

1 Work in pairs.

Student A: Turn to Communication activity 28 on page 97.

Student B: Turn to Communication activity 35 on page 99.

2 Write a description of a room in your house.

In my bedroom there are some cupboards . . .

25 I usually have a party

Present simple: adverbs of frequency

VOCABULARY AND SOUNDS

1 Put the months in the right order.

April   August   December   February   January   July
June   March   May   November   October   September

2 Listen and check 1. As you listen, repeat the words.

3 Match.

first   second   third   fourth   fifth   sixth   seventh
eight   ninth   tenth   eleventh   twelfth

9th   7th   3rd   6th   10th   12th   1st
5th   2nd   8th   11th   4th

4 Say these numbers.

13th   14th.. 15th   16th   17th   18th   19th   20th
21st   22nd   23rd   24th   25th   26th   27th
28th   29th   30th   31st

Now listen and repeat.

5 Listen and repeat.

the first of January   the fourth of April
the second of February   the fifth of May
the third of March   the sixth of June

6 Say these dates.

7th July   8th August   9th September   10th October
11th November   12th December

Now listen and check.

7 Complete.

1 My birthday is on ___.

2 My friend's birthday is on ___.

3 My mother's birthday is on ___.

4 My father's birthday is on ___.

READING AND SPEAKING

Read and find answers to the questions.

1 What do you usually do on your birthday?

2 Who do you usually spend your birthday with?

3 Do you usually get presents and birthday cards?

4 Which birthdays are always very special?

'We don't often do very much. I usually invite my friends to a bar for a drink after work. We have a cup of coffee or a drink, and then we go home.' Pablo, Spain

'Children always have a party at home. Their friends bring presents and we play games and then we have something to eat and drink. Then everyone always sings Happy Birthday.' Alexis, England

'I don't often do anything special. I sometimes go to the theatre with my wife or for a meal in a restaurant.' Dave, USA

'For us, every twelfth year of life is special, and there's usually a party. The sixtieth birthday is always very special. We usually give presents of fruit, flowers and cakes.' Kanda, Thailand

GRAMMAR

> **Present simple: adverbs of frequency**
> I **always** have a party.   100%
> I **usually** go out with friends.
> I **often** go to a restaurant.
> I **sometimes** invite friends home.
> I **never** do anything special.   0%

1 Answer the question.

Where do you put the adverb of frequency?

2 Decide where the adverbs in brackets go.

1 He goes out with friends. (often)

2 They have a party. (never)

3 She sees her parents. (usually)

4 Do you go out to a restaurant? (often)

5 I have a drink with some friends. (sometimes)

6 I get some cards and presents. (usually)

LISTENING AND SPEAKING

1 Look at the statements in the chart below. Tick (✔) the statements which are true for you or correct them with a suitable adverb of frequency.

2 🖭 Listen to Karen, Peter and Molly and tick the statements which are true for them.

3 Work in pairs and check your answers to 2.

Karen always has a party.

4 Work in pairs and complete the *Your partner* column.

What do you do on your birthday?
I usually go out with friends.

5 Find people with a birthday:

in the same month.

on the same day.

What do they do?

	You	Karen	Pete	Molly	Your partner
I always get presents and birthday cards.					
I usually go out with friends.					
I often go to a restaurant.					
I sometimes invite friends home.					
I always have a meal with my family.					
I usually have a party.					
I never do anything special.					

26 | *I can cook*

can for ability

PARENTS NEED HELP!

Do you like children? Do you like to travel? Can you speak English? Can you play the guitar? Have you got six weeks in July and August?
We need someone to go on holiday with us to the USA and help us with our three children.
Write to Mr and Mrs Burroughs, 11, Belsize Park, London.

FLAT SHARE

Three students need a fourth who can sing, do the shopping, ride a bicycle and cook to share a flat in Finchley.
£250 per month. Write to 4, Kings Close, Bromley

INTERNATIONAL SUMMER SCHOOL ASSISTANT needed.

Can you swim, play the piano, speak any languages, play tennis and football? You can? Do you need a holiday job?

Great! Ring Dave on 27566.

SECRETARY needed.

We need someone to use a computer and understand French for three mornings/week. Some weekend work. Ring 099875

VOCABULARY AND SOUNDS

1 🔲 Listen and repeat.

| dance swim draw cook sing drive type |

2 Turn to Communication activity 30 on page 97.

3 Match the words in the two boxes.

| play ride use speak understand |

| piano bicycle football computer guitar French
English Italian tennis |

play the piano

Now look at the adverts and check your answers.

4 Look at the cartoon and check you understand these words.

| need help |

5 🔲 Listen and repeat.

/æ/ Can you swim?   Yes, I can.
     Can you type?   No, I can't.

/æ/ Can you use a computer?   No, I can't.

/ə/ I can swim.   I can play the piano.
     I can't use a computer.

READING AND LISTENING

1 Read the adverts in *Do you need help?* Make notes on what each person needs to do.

2 Read the interview. Which advert is it for?

INTERVIEWER	So, are you a student?
FRANK	____
INTERVIEWER	And you need a holiday job?
FRANK	____.
INTERVIEWER	Can you swim?
FRANK	____.
INTERVIEWER	And music? Can you play the piano?
FRANK	____. But I can play the guitar.
INTERVIEWER	OK, and can you speak any languages?
FRANK	____. I can speak French and German.
INTERVIEWER	And what about sport? Can you play tennis and football?
FRANK	____.
INTERVIEWER	Good.

3 Work in pairs and guess what Frank says.

[cassette] Now look at the notes you made in 1 and listen. Put a tick (✔) by the things he can do.

GRAMMAR

Can		
Can you swim?	*Yes, I can.*
Can you play the piano?	*No, I can't.*
Can he swim?	*Yes, he can.*
Can he play the piano?	*No, he can't.*
I can swim. He can dance.	*She can cook.*
I can't cook. He can't drive.	*She can't play the piano.*

1 Work in pairs. Act out the interview in *Reading and listening* activity 2.

2 Say what Frank can and can't do.

He can swim, but he can't play the piano.

3 Work in pairs. Act out the interview again. Give true answers. Change round when you're ready.

4 Work in pairs. Look at the other adverts. Which things can you do?

I can speak English, but I can't play the guitar.

5 Work in new pairs. Tell each other what your old partner can and can't do.

Fabrice can speak English, but he can't play the guitar.

LISTENING AND SPEAKING

1 [cassette] Listen to two more interviews, with Janie and Lois, for the job. Write what each person can do.

Janie - swim, play the piano

2 Work in pairs. Who gets the job?

3 Work in groups of four. Choose one of the adverts.

Student A: You need help. Write questions you want to ask Students B, C and D.

Students B, C and D: You need a job. Working alone, think about what you can do for the job.

4 Act out job interviews in your groups.

27 *Can I have a sandwich, please?*

Talking about food and drink

VOCABULARY AND SOUNDS

1 🔲 Listen and repeat.

coffee orange milk water bread
cake apple pie cheese tomato lettuce

2 Work in pairs. Look at the picture. Point and say.

3 🔲 Listen and repeat.

cup glass bottle piece

4 🔲 Listen and repeat.

/ə/ a cup of coffee
a cup of tea
a glass of wine
a glass of milk
a bottle of wine
a bottle of water
a piece of cake
a piece of cheese
a piece of pie

Now point and say.

LISTENING

Halley Court Café Menu

Sandwiches £2.50

Cheese
Cheese and tomato
Chicken and lettuce
Beef
Halley Court Special

Pizzas £3.00

Cheese and tomato

Drinks 75p

Cola
Fanta
Milk
Mineral water
Coffee
Tea

Baked potatoes £2.00

filled with
Cheese
Chilli
Tuna and Mayonnaise

Pasta £3.00

Spaghetti

Desserts £1.00

Chocolate cake
Apple pie
Yoghurt (50p)

1 Decide where these sentences go.

WAITER Can I help you?

JANE   (1) _____

WAITER It's a sandwich with chicken, lettuce, tomato and mayonnaise.

JANE   (2) _____

WAITER It's £2.50.

JANE   (3) _____

WAITER Certainly. And anything to drink?

JANE   (4) _____

WAITER OK, a Halley Court Special and a cup of coffee. Anything else?

JANE   (5) _____

WAITER Thank you, . . . OK, a Halley Court Special, a cup of coffee and a piece of chocolate cake. Here you are.

JANE   (6) _____

WAITER Enjoy your meal.

a A piece of chocolate cake, please.

b OK, can I have a Halley Court Special, please?

c Thank you very much.

d A cup of coffee, please.

e Yes, what's a Halley Court Special?

f How much is it?

2 Work in pairs and check your answers to 1.
  Now listen and check.

FUNCTIONS

> **Talking about food and drink**
>
> *What's a Halley Court Special?*   *How much is it?*
> *Can I have a sandwich?*   *Can I have a cup of coffee?*
> *Anything to drink?*   *Anything to eat?*
> *Anything else?*   *Enjoy your meal.*
> *Here you are.*
> *Thank you.*

1 Look at the expressions in the functions box. Who says them, *waiter* or *customer*?

Waiter   *Anything to drink?*

Customer   *What's a Halley Court Special?*

2 Work in pairs. Act out the conversation in *Listening* activity 1.

LISTENING AND SPEAKING

1 Make a list of food and drink words from this lesson and from Lesson 22.

2 Listen to Finn and Selina talking about food and drink. Use the list from 1 and tick (✔) the things they like and cross (✗) the things they don't like. Add any items they like or don't like.

3 Work in pairs. Guess what they choose from the menu in *Listening* activity 1.
  Now listen and check.

4 Work in pairs. Act out conversations in the café. Use the menu in *Listening* activity 1. Ask and say.

Can I help you?
Yes, can I have a baked potato?

5 Write a typical restaurant menu for your country.

Pasta ...

Pizza ...

6 Work in groups of four or five.

Student A: You're a waiter. Act out a conversation with Students B, C and D. Use the menu you wrote in 5.

Students B, C and D: Look at Student A's menu. Act out a conversation in a restaurant.

Can I help you?

Yes. I'd like some pasta, please.

28 *Where's the station?*

Asking for and giving directions

VOCABULARY AND SOUNDS

1 [🔊] Listen and repeat.

> bank pub baker cinema chemist
> market station restaurant library
> bookshop car park post office

2 Use some of the words in 1 and complete the map.

3 Work in pairs.

Student A: Turn to Communication activity 31 on page 98.

Student B: Turn to Communication activity 41 on page 101.

Now complete the rest of the map.

4 Match the words and expressions with the signs.

> turn left go straight ahead turn right

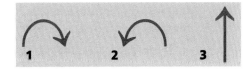

LISTENING

1 Match the question and answers.

1 Where's the station?
2 Where's the bookshop?
3 Where's the market?
4 Where's the chemist?

a It's in North Street.
b It's in West Street.
c It's in South Street.
d It's in East Street.

[🔊] Now listen and check

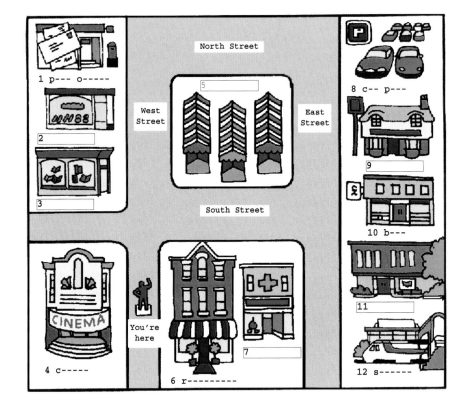

2 You're at the point marked YOU'RE HERE. Match the questions and answers.

1 Where's North Street?
2 Where's West Street?
3 Where's South Street?
4 Where's East Street?

a Turn right.
b Go along West Street. Turn right into North Street.
c Turn right into South Street. Turn left into East Street
d Go straight ahead.

[🔊] Now listen and check.

3 [🔊] Listen to four people and follow their routes. Say where they want to go.

First person - post office

4 Work in pairs and check your answers.

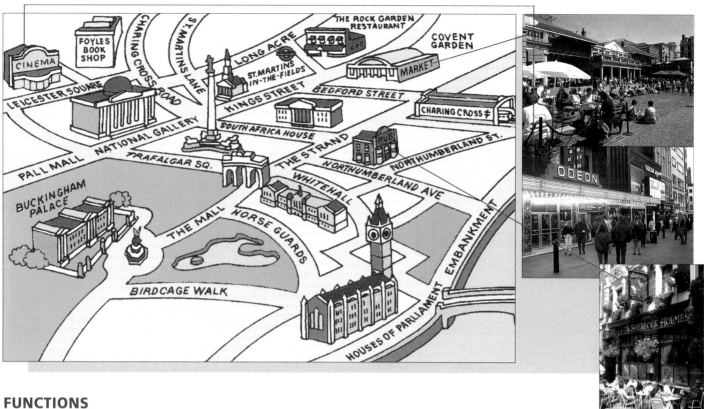

FUNCTIONS

Asking for and giving directions

Where's the station?	*It's in East Street.*
Where's West Street?	*Go along South Street.*
	Turn left. Turn right.
	It's on the left. It's on the right.
	It's straight ahead.

1 Complete.

Where's the post office?

_____ *in West Street.*

_____ *West Street?*

Go ____ South Street and _____ right. It's on the ____.

2 Look at the map on this page. Write sentences and say where these places are.

Sherlock Holmes pub   Charing Cross Station
Odeon cinema   Covent Garden market
Foyles bookshop

3 Work in pairs. Decide where you are on the map. Now write directions to the places in 2.

Where's Charing Cross Station?

Go along Pall Mall. Turn right into Trafalgar Square. Turn left into the Strand. Go straight ahead and it's on the right.

LISTENING AND READING

1 🔲 You're tourists on a walking tour of London. Listen and point to where you are. Now point to these places:
– Buckingham Palace   – The Houses of Parliament.

2 Read *Walking tour of Covent Garden and Trafalgar Square* and point.

Walking tour of Covent Garden and Trafalgar Square

Start your tour at Trafalgar Square. Go along Northumberland Avenue and turn left into Northumberland Street, where the Sherlock Holmes pub is. Turn left into the Strand and go past Charing Cross Station. Go along the Strand and turn left into Bedford Street. The old Covent Garden Market is on the right.

WRITING

Write a walking tour of your city/town.

Start your tour at the station. ...

He's buying lunch

Present continuous (1)

VOCABULARY AND READING

1 Match the sentences with the photos.

| buy | sit | run | drive | stand |

a He's buying lunch.

b They're sitting in a theatre.

c They're running.

d They're standing in a queue.

e They're driving to work.

2 Match the clocks to the photos.

3 Say where you think the people in the photos are.

A In London at the moment, people are stopping work and leaving their offices. They're standing in queues for buses, running for trains or walking to pubs or cafés and having something to eat.

B In New York at the moment, people are having something to eat and drink, such as a sandwich and coffee. They're seeing friends and having lunch or shopping.

C In Moscow at the moment, some people are having dinner in restaurants, going to the theatre or having a coffee in bars. Many families are at home. They're watching television, reading the newspaper or playing games.

D In Hong Kong at the moment, most people are lying in bed asleep, but some people are standing in queues outside clubs, or are inside and drinking or dancing.

E In Los Angeles at the moment, people are getting up, washing, getting dressed and having breakfast or driving to work.

GRAMMAR

> ### Present continuous (1)
> **Present participle form**
> *play – playing   drink – drinking*
> *stand – standing   run – running*
> *shop – shopping   lie – lying   sit – sitting*
>
> *I'm playing tennis.*   (= I am)
> *You're standing in a queue.* (= you are)
> *We're driving to work.*   (= we are)
> *They're sitting in a bar.*   (= they are)
> *He's buying a newspaper.*   (= he is)
> *She's running.*   (= she is)
> *It's stopping.*   (= it is)

1 Complete the sentences.

You form the present continuous with the verb ____ and the present participle. You use ____ with *I*. You use ____ with *we, you* and *they*. You use ____ with *he, she* and *it*.

2 Look at the passages in *Vocabulary and reading* activity 3 and write the *-ing* form of the following verbs.

> dance drink get dressed drive get up
> have lie play read run see shop
> stop walk wash watch

3 Look at the verbs in 2. Put them in groups.

Verbs ending in *-e*: *dance*

Verbs ending in two vowels: *lie*

Verbs ending in *n, p, t*: *run*

Now write the present participle with these verbs.

4 Say what people are doing at the moment.

1 My father/mother _____.

2 My sister/brother _____.

3 My friend _____.

4 My teacher _____.

5 I _____.

6 We _____.

1 My mother is going to work .

SOUNDS

1 🔲 Listen and tick (✔) the correct words.

1 a stand in   b standing   5 a play in   b playing

2 a read in   b reading   6 a run in   b running

3 a shop in   b shopping in   7 a lie in   b lying

4 a sit in   b sitting   8 a write in b writing

2 🔲 Listen and repeat the words in 1.

SPEAKING

1 Work in pairs. Turn to Communication activity 32 on page 98.

2 Work in pairs. Say

– what the time is at the moment

It's 3 pm.

– what people are doing at the moment in your country

In my country, people are ...

– what people are doing in London, New York, Moscow, Hong Kong and Los Angeles at the moment.

In London, people are ....

30 *He isn't having a bath*

Present continuous (2): negatives; questions

LISTENING AND VOCABULARY

1 Match the *wh-* questions with the correct picture.

a What's he doing?

b What's she doing?

c What are they doing?

Now work in pairs and check your answers.

2 Match the *yes/no* questions with the correct picture.

d Is she talking to her daughter?

e Is she making tea?

f Is he listening to the radio?

g Is he having a bath?

h Are they waiting for a bus?

i Are they having dinner?

3 🔲 Listen and check your answers to 1 and 2.

Picture 1 What's he doing?

Is he listening to the radio?

4 Match.

| talk   wait   listen |

for   to

5 Match.

| make   have |

coffee   tea   lunch   dinner
a bath   shower

6 🔲 Listen and tick (✔) the right sentences.

1 a He's listening to the radio. ☐   b He's listening to the CD. ☐

2 a She's making tea. ☐   b She's making coffee. ☐

3 a They're waiting for a bus. ☐   b They're waiting for a taxi. ☐

4 a He's having a bath. ☐   b He's having a shower. ☐

5 a She's talking to her son. ☐   b She's talking to her daughter. ☐

6 a They're having dinner. ☐   b They're having lunch. ☐

4 Put the adverbs

1 I go swimmir

2 We go to the

3 They go to a

4 He goes out

5 She invites h

6 You have a s

5 Write sentences

1 Can he swim

2 Can he drive

3 Can he use a

4 Can he play

5 Can he cook

6 Can he speak

6 Correct these se
doing at the m

1 They're singi

2 She's playing

3 He's running

4 They're play

5 He's eating.

6 She's washin

They aren't sin

GRAMMAR

Present continuous (2): negatives

I'm not having	*(= am not)*
he isn't having	*(= is not)*
she isn't having	
you aren't having	*(≠ are not)*
we aren't having	
they aren't having	

Questions

Are you having?	*Yes, I am.*
	No, I'm not.
Are they having?	*Yes, they are.*
	No, they aren't.
Is he/she having?	*Yes, he/she is.*
	No, he/she isn't.
What are you doing?	*What are they doing?*
What is he doing?	*What is she doing?*

SOUNDS

1 🔲 Listen an

/aɪ/ wine knif

/ɪ/ drink swi

/uː/ spoon ro

/iː/ eat read

/ʊ/ book coo

2 🔲 Listen ar

bicycle veget

America Dece

Now say the w

1 Write answers to the questions in *Listening and vocabulary*, activity 2.

Is he listening to the radio?

No, he isn't. He's listening to a CD.

2 Tick (✔) or correct the sentences.

At the moment, ...

1 ... my teacher is dancing.

2 ... my friends are playing badminton.

3 ... my teacher is wearing a yellow hat.

4 ... I'm reading the newspaper.

5 ... I'm lying on the beach.

6 ... we're having breakfast.

My teacher isn't dancing. She's working.

3 Write questions.

1 you are speaking English

2 your teacher is reading

3 your friends are listening to the teacher

4 your brother is going to work/school

5 your mother/wife/husband is waiting outside

6 you are writing questions

Are you speaking English?

4 Answer the questions you wrote in 3.

SPEAKING AND WRITING

1 Work in pairs. Turn to Communication activity 33 on page 98.

2 Write down as many differences between the pictures as you can in 1 minute.

34 | *Was she in the kitchen?*

Past simple (2): *yes/no* questions and short answers

VOCABU

1 Look at
a car th
the bedr
Look at
draw s
Now loc
blue la
These w
I you
These w
in at
These w
always
Work ir
or adjec

noun v

white
unders

2 Here ar
part of
Does sl
I can c
What d
Can I h
I don't
Where'
There's
He's bu
I usual
He isn'

VOCABULARY

1 Complete these sentences with *food* and *drink*.

> hungry   thirsty

When you're hungry, you need some ____.

When you're thirsty, you need a ____.

2 Here are some new words in the story in this lesson. Check you understand them.

professor colonel detective sergeant lady
Miss Mrs butler murder knife murderer
scream dead cook alone

What kind of story do you think it is?

READING AND LISTENING

1 Match the drawings and the paragraphs.

A It was eight o'clock in the evening at Ripley Grange, the home of Lady Scarlet. There were six people at the table in the dining room: Colonel White, Miss Green, Professor Peacock, Mrs Mustard, Doctor Plum, and Lady Scarlet.

B It was cold in the dining room. They were hungry because the food wasn't very good. There was a cold potato, a piece of cold meat and a piece of cold lettuce.

C They were thirsty, but there wasn't any wine, only a glass of water.

D Probe, Lady Scarlet's butler and cook was at the door of the dining room. He was tall with black hair but he wasn't very good-looking. He wasn't a very good butler, and he was an awful cook.

🔊 Now listen and check.

2 Look at these sentences from the next part of the story. Put them in the correct order.

☐ Then, there was a scream. It was Lady Scarlet. Was she in the kitchen? Yes, she was.

☐ Was there something on the table? Yes, there was – there was a knife on the table. It was red. Was it murder? Who was the murderer?

☐ It was ten o'clock in the evening at Ripley Grange. It was quiet ... very quiet. Was there anyone in the dining room? No, there wasn't.

☐ Was she alone in the kitchen? No, she wasn't. Probe was also in the kitchen. He was dead.

🔊 Now listen and check.

GRAMMAR

Past simple (2): *yes/no* questions and short answers	
Was I in the kitchen?	*Yes, you were.*
	No, you weren't.
Were you in the kitchen?	*Yes, I was.*
	No, I wasn't.
Was he/she in the kitchen?	*Yes, he/she was.*
	No, he/she wasn't.
Were we in the kitchen?	*Yes, we were.*
	No, we weren't.
Were they in the kitchen?	*Yes, they were.*
	No, they weren't.

Put the words in order and write questions.

1 was Colonel White the kitchen in?

2 Mrs Mustard was in the dining room?

3 Miss Green in the bedroom was?

4 was Doctor Plum the living room in?

5 Professor Peacock in the garden was?

6 Lady Scarlet was in the kitchen?

LISTENING AND SPEAKING

1 🔲 Detective Prune and Sergeant Peach come to Ripley Grange next morning. Listen to their interviews and find out where people were at 8 pm and 10 pm. Put a tick (✔) in the chart below.

2 Work in pairs and check your answers.

Was Colonel White in the kitchen at ten o'clock?

No, he wasn't. He was in the living room.

3 Complete the sentences. Say where people were at 10 pm

Colonel White was in the ____.

Miss Green was in the ____.

Professor Peacock was in the ____.

Mrs Mustard was in the ____.

Doctor Plum was in the ____.

Lady Scarlet was in the ____.

4 Work in pairs. Who was the murderer?

5 🔲 Listen and find out who was the murderer, and why.

Colonel White

Miss Green

Professor Peacock

Mrs Mustard

Doctor Plum

Lady Scarlet

kitchen

bathroom

bedroom

dining room

living room

garden

35 *They didn't have any computers*

Past simple (3): *had*

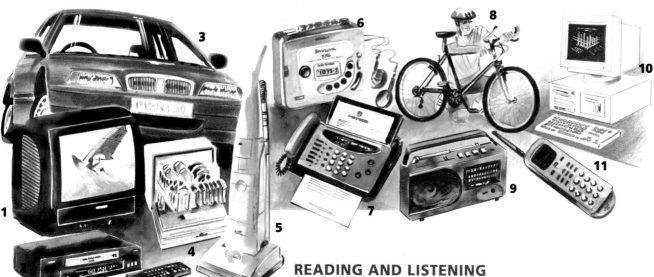

VOCABULARY AND SOUNDS

1 Match the words with the pictures.

> television   video recorder   telephone
> radio   computer   personal stereo   car
> bicycle   fax machine   dishwasher
> vacuum cleaner

2 Tick (✔) the things you or your family have at home.

3 🔊 Listen and underline the stressed syllables in the words in 1.

> *tele**vi**sion*

🔊 Now listen and repeat.

4 Work in pairs. Ask and say which things in 1 you have at home.

> *We've got a television and a video recorder.*

5 Circle the things your grandparents had.

READING AND LISTENING

1 Mary is 80. Read about her and her family. Tick (✔) the things they had and put a cross (✗) by the things they didn't have when she was a child.

MARY	Well, when I was a child, we had a telephone in the hall, but we didn't have a television. No one had television. And, of course, we didn't have a video recorder.
INTERVIEWER	What about a radio?
MARY	Yes, we had a radio. The radio programmes were very good.
INTERVIEWER	And computers, did you have computers or personal stereos?
MARY	Oh, no, we didn't have computers or personal stereos. And I was fifty before I had my own radio.
INTERVIEWER	Did you have a car?
MARY	Yes, some families had a car, but we didn't. We had bicycles, all six of us.
INTERVIEWER	And no fax machine, of course. Or vacuum cleaner or dishwasher?
MARY	No, we didn't have a fax machine. But we had a kind of vacuum cleaner. And we didn't have a dishwasher.

telephone		
television		
video recorder		
radio		
computer		
personal stereo		
car		
bicycle		
fax machine		
vacuum cleaner		
dishwasher		

2 Work in pairs and check your answers.

3 🔊 Listen to Mary and put a \ if you hear any extra words or phrases.

4 Work in pairs and check your answers. Try to remember the extra words or phrases.
🔊 Now listen again and check.

GRAMMAR

> Past simple (3): *had*
>
> | I | | I | |
> |---|---|---|---|
> | you | | you | |
> | he | | he | |
> | she | had | she | **didn't** have (= did not have) |
> | it | | it | |
> | we | | we | |
> | they | | they | |
>
> We **had** a telephone in the hall, but we **didn't have** a television.

1 Write sentences saying what Mary's family had or didn't have.

They had a telephone, but they didn't have a television.

2 Write sentences saying what your grandparents had or didn't have.

My grandparents had a television, but they didn't have a personal stereo.

3 Complete with *was, were* or *had*.

1 It ____ a great film.
2 You ____ breakfast at seven o'clock.
3 They ____ dinner at eight o'clock.
4 The play ____ awful. We ____ bored.
5 We ____ a holiday in August.
6 He ____ thirsty so he ____ a drink.

SPEAKING AND LISTENING

1 Work in pairs. Turn to Communication activity 38 on page 100.

2 Work in pairs and check your answers to 1.

They didn't have telephones.

3 Here are some more discoveries and inventions which are now everyday items.

aeroplanes trains cameras newspapers steam engines

4 🔊 Listen and repeat these dates.

1950 1850 1750 1650 1550

5 Work in pairs. You're going to play *When in the World ...?* Ask and say:

Did they have trains in 1850? Yes, they did.
Did they have trains in 1550? No, they didn't.

6 🔊 Listen to two people playing *When in the World ...?* and check your answers. Score one point for each correct answer.

7 Work in groups of three or four. Talk about things you had or didn't have when you were a child.

36 *We listened to the radio*

Past simple (4): regular verbs

VOCABULARY

1 Write these words in the correct column.

last month   this year   this month   today
this week   last week   last year   yesterday

In the past   In the present
last year   *this year*

2 Read.

This year is 1997. Last year was 1996.

3 Complete.

1 This year is ____. Last year was ___.

2 This month is ____. Last month was ____.

3 Today is ____. Yesterday was ____.

READING AND LISTENING

1 Read about Brian's childhood. Match the headings
and the paragraphs. There is one extra heading.

Holiday   Sunday   Work   Sport
Entertainment at home

A 'We had a television set and we watched it in the
evening. We listened to the radio or played records.
There weren't any computers or computer games, but
sometimes my brother played the piano and my sister
and I danced to the music. In the summer, we played
games in the garden until eight o'clock.'

B 'When I lived in London with my parents, I played
tennis a lot, but I don't play it now. I wanted to play
football, but I was terrible. On Sunday mornings, I
watched the local football team in the park. I started
running when I was at university in 1973.'

C 'On Sunday my mother cooked Sunday lunch and we
had Sunday lunch together. Then in the afternoon,
we had a walk in the country or I visited my friends.
In summer we watched the village cricket match. We
stayed at home on Sunday evenings.'

D 'My father worked five days a week. He started at 9 am
and finished at 5 pm. He worked in a bank in London.
We lived in South London, so every day he travelled
over two hours by train to the city centre to get to
work and back.'

2 Correct the sentences.

1 His sister played the piano.

2 He started running at school.

3 He visited his friends on Sunday evening.

4 His father worked in a shop.

5 They lived in the country.

6 They watched television in the morning.

7 He had Sunday lunch with his friends.

8 His father walked to work.

GRAMMAR

> **Past simple (4): regular verbs**
> /d/ **played stayed**
> *He played tennis.*   *They stayed at home.*
> /t/ **worked watched**
> *My father worked in London.*   *He watched the cricket.*
> /ɪd/ **wanted visited**
> *He wanted to play football.*   *He visited her friends.*

1 Complete the sentence.

You form the past simple of most regular verbs by adding _____.

2 Look for the past simple form of these verbs in *Reading and listening* activity 1, and write them down.

1 listen play want start cook visit
  watch stay work

2 dance live

3 travel

SOUNDS

1 [🔊] Listen and repeat.

/d/ played stayed

/t/ worked finished

/ɪd/ wanted visited

2 Write these words in the correct column.

liked lived arrived cooked stopped danced
started opened hated listened helped changed

/d/   /ɪd/   /t/

[🔊] Now listen and check your answers.

LISTENING AND WRITING

1 [🔊] Listen to two more people, Ben and Judy, talking about when they were children. Put the letter corresponding to what they say in the correct column of the chart.

	Ben	Judy
Entertainment at home		
Sport		
Sunday		
Holiday		
Work		

a 'We stayed in a small hotel in Devon.'

b 'We travelled around France.'

c 'I played tennis in my school team.'

d 'I played football with friends.'

e 'We often had friends for lunch.'

f 'My father was an engineer and he travelled a lot.'

g 'My father and my mother were doctors.'

h 'We played cards.'

i 'My mother cooked lunch.'

j 'I played the piano.'

2 Work in pairs and check your answers.

[🔊] Now listen again.

3 Write five sentences about yourself ten years ago. Use some of these verbs.

live start play cook visit stay work finish
watch listen dance

4 Work in pairs. Show your sentences to each other. Tell the rest of the class what your partner wrote.

37 *Picasso didn't live in Spain*

Past simple (5): negatives

VOCABULARY AND READING

1 Check the verbs you don't know in a dictionary.

be born   start   study   live   work
paint   finish   die

2 Write the past simple of the verbs in 1.

3 Complete the sentences with the verbs in the past simple.

1 I ____ work at 9 am yesterday morning.

2 She ____ medicine at university.

3 We ____ in Paris for ten years.

4 Picasso ____ Guernica.

5 He ____ in a bank last year.

6 She ____ work at five yesterday afternoon.

7 Shakespeare ____ in 1564 and ____ in 1616.

1

4 Match these words with the photos.

sculpture   painting

5 Work in pairs. You're going to read about Pablo Picasso. Do you know:

– who he was?

– when and where he was born?

– where he lived?

– what he did?

– where and when he died?

6 Read and check your answers to 5.

7 Tick (✔) the true statements.

1 Picasso was Spanish.

2 He was born in 1881.

3 He started to paint when he was fifteen.

4 He lived in Barcelona in 1904.

5 He worked in Madrid from 1904.

6 He finished Guernica in 1938.

7 He returned to Spain to live.

8 He died in 1972.

PICASSO, Pablo (1881 – 1973)

Pablo Picasso was a great Spanish artist. He made over 20,000 sculptures and paintings in his life. He was born in Malaga in 1881, and started to paint when he was ten. When he was fifteen, he studied at the Barcelona School of Fine Arts. From 1904 he lived and worked in Paris. Guernica was his most famous painting. He painted it in 1937 and finished it in two months. He didn't want the painting to be in Spain so it was in a gallery in New York for many years. You can see it now in the Prado Museum, in Madrid. For most of his life he didn't live in Spain. He lived in France and died there in 1973.

2

GRAMMAR

Past simple (5): negatives

I
you
he
she   *didn't live*   (= did not)
it
we
they

He **didn't** *want the painting to be in Spain.*
He **didn't** *live in Spain.*

---

1 Complete.

To form the past negative you write
_____ before the infinitive of the verb.

2 Correct the false statements in
Vocabulary and reading activity 7.

*3 He didn't start to paint when he was
fifteen. He started to paint when he
was ten.*

3 Write things you did and didn't do:

– yesterday   – last week

– last month   – last year

Yesterday I played football.

I didn't listen to the radio.

SOUNDS

1 🔊 Listen and repeat.

He <u>didn't</u> start to paint when he was <u>fifteen.</u> He <u>started</u> to <u>paint</u> when he was <u>ten.</u>

He <u>didn't</u> work in <u>Madrid</u> from 1904. He <u>worked</u> in <u>Paris.</u>

He <u>didn't</u> finish <u>Guernica</u> in <u>1938.</u> He <u>finished</u> it in <u>1937.</u>

2 Underline the stressed words in your other answers to *Grammar* activity 2.

🔊 Now listen and check. As you listen say the sentences aloud.

SPEAKING AND WRITING

1 Work in pairs. Think about the life of a famous person, and make notes on:

– when he or she was born   – where he or she lived

– what he or she did   – where and when he or she died

Atatürk, Kemal – born in 1881

2 Work in groups of three or four. Describe the famous person you made notes on in 1. Don't say his or her name. The others should guess who he/she is.

3 Write a paragraph about the famous person.

Kemal Atatürk was born in 1881.

4 Write two true sentences about your life, and one false sentence.

I was born in Italy.

I lived in Milan and Florence.

I painted the Last Supper.

5 Work in pairs. Show your partner your sentences. Can he or she guess which is the false one?

You didn't paint the Last Supper!

38 *Did you take a photograph?*

Past simple (6): *yes/no* questions and short answers

VOCABULARY

You are going to read three stories in this lesson.
Check these words in the dictionary.

| weather foggy dark bark village |
| disappear dog cat climb |

READING AND LISTENING

1 Read the conversation and decide where these
sentences go.

a No, we didn't.   c No, we didn't.

b Yes, he did.   d No, he didn't.

STEVE   When we started, the weather was fine.
We walked about fifteen kilometres.
About four hours later, we started to come
home. Suddenly, the weather changed. It
was very foggy. We were still in the
mountains.

PHILIP   Did you stay there?

STEVE   (1) ____ . We walked for an hour but it
was very cold and dark. We decided to
stop walking and wait.

PHILIP   Did you wait for a long time?

STEVE   Well, no, we didn't. You see, after about
two or three minutes, something
happened. A dog, a big black dog
suddenly appeared out of the fog. He had
red eyes. Then he barked at us.

PHILIP   Did he want to help you?

STEVE   (2) ____ . So we walked for two hours
behind the dog. Then we arrived back at
the village.

PHILIP   Did the dog stay with you?

STEVE   (3) ____ . When we arrived at the village,
he disappeared into the mountains.

PHILIP   Did you take a photograph?

STEVE   (4) ____ .

2 Work in pairs and check your answers.

 Now listen and check.

SOUNDS

 Listen and read.

1 Did you stay there?

2 Did he want to help you?

3 Did the dog stay with you?

4 Did you take a photograph?

 Now listen and repeat the sentences.

GRAMMAR

> **Past simple (6):** *yes/no* questions and short answers
>
> | | | | | |
> |---|---|---|---|---|
> | | *I* | | *Yes, you did.* | *No, you didn't.* |
> | | *you* | | *Yes, I did.* | *No, I didn't.* |
> | | *he* | | *Yes, he did.* | *No, he didn't.* |
> | **Did** | *she* | *take a photo?* | *Yes, she did.* | *No, she didn't.* |
> | | *we* | | *Yes, we did.* | *No, we didn't.* |
> | | *they* | | *Yes, they did.* | *No, they didn't.* |
>
> *Did you take a photo?   No, we didn't.*

1 Complete this sentence.

You form *yes/no* questions with ____ + subject
(*I, you, he* etc.) + ____.

2 Match questions and answers.

1 Did you follow the dog?   a Yes, you did.
2 Did she get lost?   b Yes, he did.
3 Did I say hello?   c No, they didn't.
4 Did they get lost?   d No, she didn't.
5 Did he want to follow him?   e Yes, I did.

3 Complete the questions and answers.

1 Did you sleep well?   Yes, ____.
2 Did she call you?   No, ____ .
3 Did they come by train?   Yes, ____.
4 ____ he say hello?   No, ____.
5 ____ I answer the question? Yes, ____.
6 ____ I take your keys?   No, ____.

4 Write the questions.

1 you go cinema last did to night the?
2 did watch you yesterday television?
3 you did see friends your last weekend?
4 week you swimming did last go ?
5 school last you go to did week?
6 today your you did bring books?

5 Work in pairs. Answer the questions in 4.

1 Did you go to the cinema last night?   No, I didn't.

READING AND WRITING

1 Read and answer the questions.

Barbara Paule lived in Pennsylvania with her cat, Muddy Water. On 23 June, 1985, she was in her car with the cat in Dayton, Ohio. The car stopped and the cat climbed out of the window and disappeared. She was very unhappy and went home to Pennsylvania. In June 1988, a cat arrived at the door of her house. She washed it and then saw it was Muddy Water. The cat came home after three years and 800 kilometres.

1 Did Barbara Paule live in Pennsylvania?
2 Did the cat climb out of a window at home?
3 Did Barbara find the cat in Dayton?
4 Did Barbara go home to Pennsylvania?
5 Did she know the cat at the door?
6 Did Muddy Water return home?

2 Read these questions which come from a story about a dog. Try to answer the questions before you read the story.

1 Was Dr Ueno a teacher at a university in Tokyo?
2 Did he walk with his dog, Hachiko, to the station every day?
3 Did Hachiko then walk home?
4 Did he return in the evening to meet Dr Ueno?
5 Did Dr Ueno arrive one day?
6 Did Hachiko wait for him?
7 Did Hachiko return to the station and wait for his friend every day?
8 Did he do this for nine years?

3 Now try to write as much of the story as you can, using your answers to the questions in 1.

Dr Ueno was a teacher at a university in Tokyo.

4 Turn to Communication activity 37 on page 99 and read the story. Correct the version you wrote in 3.

Past simple (7): questions; irregular verbs

Brochure
Visit the
Statue of
Liberty

VOCABULARY AND SOUNDS

1 🔊 Listen and repeat.

| bill   tickets   receipts   brochure |

2 Work in pairs. Point and say.

3 Write the present tense of these verbs.

drank   spent   got   left   flew   saw
took   ate   sat   went   bought   thought

drink …

Now work in pairs and check your
answers.

4 Work in pairs. Check the meaning of
the verb in the dictionary.

5 🔊 Listen and repeat the verbs in 3.

JOE ALLEN
RESTAURANT BILL

Drinks $23.00
Food $40.00

TOTAL $63.00

Bloomingdale's

RECEIPT
Lady's Jacket $70.00

TOTAL $70.00

Bloomingdale's

RECEIPT
Man's Seiko Watch $150.00

TOTAL $150.00

FEB 9TH   MAJESTIC THEATRE
8.00 PM   247 W. 44TH ST. BROADWAY

$40.00   The Phantom of the Opera

FEB
8.00   CIRCLE   Saturday February 9th
       E41   8.00 PM

$40

CIRCLE   Saturday February 9th   E4
E40   8.00 PM
                                                 $40.0

Jo's Taxis

Receipt

TO: *Broadway (Majestic Theatre)* | *$30.00*

ISSUED BY **FLYAWAYS** LONDON

RESTRICTIONS/ENDORSEMENTS   SOTO   09990702 04JUN6
APEX FARE-NON REF.   RB2DAU 9350106 FLYAWAYS
NONENDORSABLE. VALID BA FLTS   NEW YORK USA
NON RETOUTABLE
PASSENGER NAME   NOT TRANSFERABLE
WALTON/DAVID MR   NO EARNED EXEC CLUB MILEAGE MKJEJ

	CARRIER	FLIGHT	DATE	TIME	STATUS	FARE BASIS	NOT VALID
FROM VOID	VOID	VOID	VOID	VOID	VOID	VOID	
TO VOID	VOID	VOID	VOID	VOID	VOID	VOID	
TO LONDON TERM 4 LGW	FA	217F	06AUG	0400	OK	LHXAN	06AUG
TO NEW YORK JFK	FA	216F	07AUG	1200	OK	LHXAN	07AUG

125 4469753068 4   FORM OF PAYMENT
                        CC CA5404382100588733-004130

LISTENING AND READING

1 🔊 Listen and read the letter.

*David and I went to New York for the weekend! We flew from
London airport and arrived in New York at 12 am on Saturday.
We took a taxi from the airport to our hotel near Central Park
and then went for lunch at Joe Allen, a famous restaurant on
Broadway. The food was great!*

*In the afternoon, we went shopping in Bloomingdale's (a very
famous shop in New York). I bought a lovely red jacket and David
got a new watch.*

*In the evening we went to a theatre on Broadway and saw The
Phantom of the Opera. A wonderful day!*

2 Write short answers to the questions.

1 Where did David and Kate go for the
   weekend?

2 When did they arrive?

3 How did they get to the hotel?

4 Where did they go for lunch?

5 What did they do in the afternoon?

6 What did they buy?

7 What did they see in the evening?

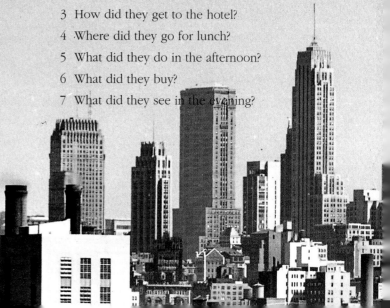

3 Look at the items in *Vocabulary and sounds*, activity 1. Write complete answers to the questions.

1 When did the plane leave London?

2 How long did the journey take?

3 How much did Kate spend on her jacket?

4 How much did David spend on his watch?

5 How did they get to the theatre?

6 Where did they sit?

7 How much were the tickets?

4 Work in pairs and check your answers to 3.

GRAMMAR

> ### Past simple (7): irregular verbs
> *fly – flew   take – took   go – went   buy – bought*
> *get – got   see – saw   drink – drank   spend – spent*
> *leave – left   eat – ate   sit – sat   think – thought*
>
> ### Questions
> *Where did they go for the weekend?*
> *When did they arrive?*
> *What did they do in the afternoon?*
> *How did they get to the hotel?*
> *How long did the journey take?*
> *How much did David spend?*

1 Write the past tense of these verbs.

go   think   buy   see   fly   take   get

went

2 Match the present and past of these verbs.

run   find   bring   do   come   give   read   hear   say
make   stand   wear

did   came   gave   read   ran   found   brought   heard
said   made   stood   wore

3 Write complete answers to *Listening and reading* activity 2.

They went to New York for the weekend.

WRITING AND SPEAKING

1 Read the next part of the letter and write questions for the missing information.

On Sunday morning we left our hotel at (1) *when* _____ in the morning and took our (2) *what* _____ to the airport. Then we went to (3) *where* _____ by boat. We spent (4) *how long* _____ there and then we returned to Manhattan. We bought some (5) *what* _____ for lunch and ate them in (6) *where* _____. We sat in the park for about (7) *how long* _____. In the afternoon, we went to the (8) *where* _____ and saw a (9) *what* _____. By now it was time to go to the airport for our flight home, so we took (10) *what* _____. We got the plane at (11) *when* _____ and flew back to (12) *where* _____. The end of a wonderful weekend!

Now work in pairs and check your answers.

1 When did they leave their hotel?

2 Work in pairs.

Student A: Turn to Communication activity 6 on page 92.

Student B: Turn to Communication activity 39 on page 100.

3 Work in pairs. Ask and answer the questions you wrote in 1 and complete the passage.

4 Write about a special weekend.

Last month I went to Paris.

40 *The end of the world?*

Tense review: present simple, present continuous, past simple

GRAMMAR

> Tense review
> **Present simple**
> *They work very hard.*
>
> **Present continuous**
> *What's happening?*
>
> **Past simple**
> *I worked all my life on the railways.*

1 Name the tense.

1 He speaks English.

2 We had dinner in a restaurant.

3 She's wearing a black dress.

4 They don't take credit cards.

5 He's writing a postcard.

6 I didn't hear you.

2 Choose the tense.

1 Ten years ago we (live) in a house in the country.

2 (Have) you dinner at the moment?

3 When (arrive) she arrive home last night?

4 What time (be) it now?

5 We usually (go) to the cinema at the weekend.

6 No, he's away at the moment. He (lie) on a beach in the South of France.

VOCABULARY

🔊 Listen and point.

> waiting room   ticket office   platform
> passenger   railway

READING AND LISTENING

1 🔊 Read and listen to part one of *The end of the world*. Answer the questions.

'My name is Joseph Finch. I live in a house in the country. My wife died a few years ago, and the children? Well, I don't see much of the children. They work very hard, they don't have much time for their father.

But, I've got a beautiful house. It's an old railway station. It's got a ticket office, a waiting room and a platform. But there aren't any trains, oh no. There weren't many passengers around here so they closed the station when the last train left thirty years ago.

I worked all my life on the railways. When I was sixty-five, I stopped work, but I still liked trains so, when I saw the old station for sale, I bought it and made it my home.

I made a beautiful garden full of flowers where, in the past, the trains ran. My kitchen was the old ticket office, my living room was the waiting room, and the platform has got a table and chair where, on summer mornings, I have breakfast. In fact, I'm sitting in the garden now. It's quiet and peaceful.'

1 Where does Joseph Finch live?

2 Does he live in a flat?

3 Is he married?

4 Has he got any children?

5 Describe his house.

6 When did they close the station?

7 What was Joseph's job?

8 When did he stop working?

9 In the past what was his kitchen?

10 What was the living room?

11 Has he got a garden?

2 Look at the next part of the story. Read the
paragraphs and put them in order.

a 'The man stood by a poster. It said, "THE END OF
THE WORLD?"'

b 'They all wore clothes from thirty years ago. "This is very
strange," I thought. I saw an old taxi and two or three old cars,
and a man selling newspapers outside the station.'

c 'Some people got out of the train and said hello to their
friends, and others got in and said goodbye.'

d 'But one day, a strange thing happened. On the night of 28
October 1992 after a cold day in my garden, I had dinner, read
my book and then I went to bed. I was asleep when suddenly
I heard a noise like a train.'

e 'I got up and looked out of the window and saw a train in the
garden. "What's happening? Am I dreaming?" I said. There were
passengers on the platform.'

3   Listen and check your answer to 1.

4 Work in pairs and answer the questions. What do you
think happened next?

1 Did Joseph Finch go back to sleep?

2 Did he get dressed?

3 Did he go downstairs?

5 Read and check.

'So I got dressed, ran downstairs and went outside. But when I
got outside, no one was there. 'Hello!' I said. But no one replied.
I walked up and down the platform. Then I went back inside. The
ticket office was my kitchen again, and the waiting room was my
dining room. The garden was full of flowers and everything was
quiet and peaceful. "It was a dream," I thought.'

6 Turn to Communication activity 42 on page 101.

7 Work in pairs. Can you think of an explanation?

Railway

Passenger

Platform

WAITING
ROOM

Progress check 31–40

1 Work in pairs. Look at all the vocabulary boxes in Reward Starter. Find:

– three words which you like

– three words which you don't like

– three words which you can use to talk about yourself

– three words which you often see in your country

– three words which sound nice

2 Tick (✔) the correct answers.

1 *September is*
   a a month ☐
   b a day of the week ☐
   c a job ☐

2 *A bank is*
   a a word to describe how you feel ☐
   b a place where you put money ☐
   c a drink ☐

3 *A car is*
   a a means of transport ☐
   b an item of clothing ☐
   c food ☐

4 *A detective is*
   a a job ☐
   b a place ☐
   c a member of the family ☐

5 *A dishwasher is*
   a a place of entertainment ☐
   b an object you see at home ☐
   c something you do on holiday ☐

6 *Mrs is*
   a a form of address for a married woman ☐
   b a form of address for a single woman ☐
   c a day of the week ☐

3 Look at the words in Lessons 31 – 40 again. Choose words which are useful to you and write them in your *Wordbank* in the Practice Book.

GRAMMAR

1 Write questions about what people are doing.

1 You/5th January?

2 Pete/5th November?

3 Jane/August?

4 Phil and Tim/this evening?

5 Ken/this weekend?

6 Julia and Philippa/this afternoon?

What are you doing on the fifth of January?

2 Answer the questions you wrote in 1.

1 fly to Thailand   4 go to the cinema

2 have a party   5 play football

3 stay with friends in Paris   6 go to an exhibition

1 We're flying to Thailand.

3 Write questions using the past simple.

1 you/thirsty?   4 they/tired?

2 he/dead?   5 she/unhappy?

3 you/hungry?   6 you/bored?

1 Were you thirsty?

4 Write answers to the questions in 3.

1 yes 2 no 3 yes 4 no 5 yes 6 no

1 Yes, I was.

5 Write sentences saying what people had or didn't have in the past.

1 televisions (✔)   5 telephones (✔)

2 videos (✗)   6 computers (✗)

3 dishwashers (✗)   7 cars (✔)

1 They had televisions. 2 They didn't have videos.

6 Write the past simple of these verbs.

play work paint be born study take
go buy see live

7 Complete the sentences with the past simple form of the verbs in 6.

1 I _____ a good film at the cinema last night.
2 He _____ in London for a year.
3 She _____ English at school.
4 Picasso _____ Guernica.
5 They're French. They _____ in Paris.

8 Write questions.

1 you/take a photo?
2 he/go home?
3 she/get lost?
4 you/go out last night?
5 they/see their friends at the weekend?
6 I/answer your question?
1 Did you take a photo?

9 Write answers to the questions you wrote in 8.

1 yes 2 no 3 yes 4 no 5 yes 6 no

10 Write questions.

1 what/you do/last weekend?
2 where/you go/on holiday last year?
3 who/you see/yesterday?
4 what/you have for breakfast/today?
5 what/you buy/at the weekend?
6 when/you get home/last night?
1 What did you do last weekend?

11 Write answers to the questions in 10.

1 I went shopping on Saturday.

SOUNDS

1 🔊 Listen and tick (✔) the word you hear.

1 hit   heat   4 sin   seen
2 sit   seat   5 bin   been
3 bit   beat   6 ill   eel

2 🔊 Listen and repeat.

/ɑ/ dog orange shop stop
/əʊ/ poster boat hotel brochure go
/ɔː/ walk hall born

LISTENING

1 🔊 You're going to hear the song *Yesterday* by the Beatles. There are some words missing in the lines below. Listen and mark a line, like this /, when you hear an extra word.

Yesterday, all my troubles seemed far away,
Now it looks as though they're here to stay,
I believe in yesterday.
Suddenly, I'm not the man I used to be,
There's a shadow over me,
Oh yesterday came suddenly
She had to go, I don't know
She wouldn't say.
I said something, now I long for yesterday.
Yesterday, love was an easy game to play,
Now I need a place to hide away,
Oh, I believe in yesterday.
Why she had to go I don't know
She wouldn't say
I said something wrong, now I long for
yesterday.
Yesterday, love was such an game to play,
Now I need a to hideaway
Oh, I believe in yesterday.

2 Work in pairs and check your answers. Can you remember what the extra words are?
🔊 Now listen again and check.

Communication activities

1 *Lesson 4*
Vocabulary and sounds, activity 5

Student A: Ask Student B to spell these words:

hello   doctor   student   seven   listen   say

Now turn back to page 10.

2 *Lesson 4*
Listening and speaking, activity 3

Student A: You're Henry Schwarzkopf and you're a doctor. Here's some more information.

Ms Fiona Pink – teacher

Now turn back to page 11.

3 *Lesson 5*
Writing and speaking, activity 2

Student A: Read the information.

Terry Crystal is an actor.
The singer is from St Petersburg.
The secretary is Italian.

Now turn back to page 13.

4 *Lesson 10*
Functions, activity 1

Student A: Look at the words for these things.

sandwiches   pizza   telephone

Now turn back to page 23.

5 *Lesson 11*
Functions and grammar, activity 3

Student A: Ask Student B how much 1-3 are and write the price. Then answer Student B's questions about 4-6.

1 jeans	4 sweater	£25.99
2 jacket	5 shirt	£30.00
3 shoes	6 skirt	£45.99

Now turn back to page 27.

6 *Lesson 39*
Writing and speaking, activity 2

Student A: Read the passage and answer the questions you wrote in 1.

On Sunday morning, we left our hotel at (1) *eight o'clock* in the morning and took our (2) - to the airport. Then, we went to the (3) *Statue of Liberty* by boat. We spent (4) - there and then returned to Manhattan. We bought some (5) *sandwiches* for lunch and ate them in (6) - . We sat in the park for about (7) *two hours*. In the afternoon, we went to the (8) - and saw a (9) *film*. By now it was time to go to the airport for our flight home, so we took a (10) -. We got the plane at (11) *midnight* and flew home to (12) -. The end of a wonderful weekend!

Now turn to page 87.

7 *Lesson 12*
Speaking and writing, activity 1

Student A: Ask Student B where these objects are and put a cross (*X*) on the picture on the right.

bag   keys   personal stereo   wallet

Now answer Student B's questions.

Now turn back to page 29.

8 *Lesson 14*

Reading and speaking, activity 2

Student A: Match the questions with the missing information, then ask and answer to complete the E-mail. If you don't know how to spell the names, ask *How do you spell ...?*

What's her name?   *What's her father's name?*

What's she like?   *What's her sister's name?*

How old is her brother?

```
Dear Brad,

Thanks for your E-mail.

My name is (1)____. I'm an English teacher and I'm
twenty-three years old. I've got (2)____ hair and (3)____
eyes. My family is from London.

My mother's name is Pippa and she's a teacher. My father's
name is (4) ____. He's a doctor. He's got blue eyes, but he
hasn't got any hair!

I've got a brother and a sister. My brother James is an
engineer, too! He's (5)____ years old and he's got fair
hair (he's very good-looking)! My sister's name is (6)____.
She's eleven; she's also got (7)____ hair and (8)____ eyes.

Write soon,

Sue
```

Now turn back to page 33.

9 *Lesson 17*

Listening and reading, activity 3

Student A: Listen and find out what time they have breakfast in Russia, lunch in Hong Kong and dinner in Mexico.

Now turn back to page 39.

10 *Lesson 5*

Writing and speaking, activity 2

Student B: Read the information.

The actor is from Los Angeles.

Mischa Godonov is Russian.

The secretary is from Naples.

Now turn back to page 13.

11 *Lesson 19*

Vocabulary and sounds, activity 3

Student A: Look at the words for these sports.

 Football

 Tennis

 Volleyball

 Table Tennis

 Skiing

Running

Now turn back to page 42.

12 *Lesson 21*

Vocabulary and sounds, activity 2

Look at the photos of different means of transport and check you know what the words mean.

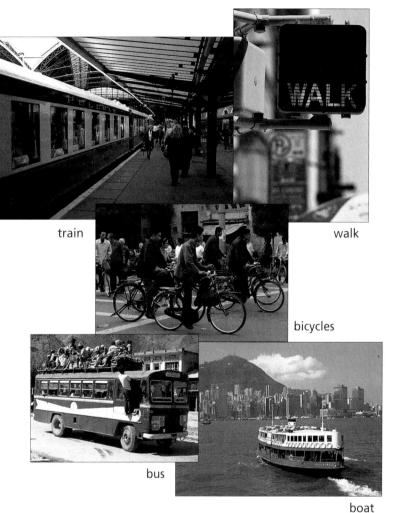

train

walk

bicycles

bus

boat

Now turn back to page 48.

13 *Lesson 22*

Vocabulary and sounds, activity 2

Student A: Look at the words for these things.

Now turn back to page 50.

14 *Lesson 22*

Grammar, activity 1

Student A: 🔊 Listen and answer.

What food do they eat in Morocco?

What do they drink in India?

What do they have for breakfast in India?

Now turn back to page 51.

15 *Lesson 4*

Vocabulary and sounds, activity 5

Student B: Ask Student A to spell these words:

goodbye   teacher   singer   eight   repeat   write

Now turn back to page 10.

tea

milk

yoghurt

rice

bread

beef

potato

beer

apple

16 *Lesson 4*

Listening and speaking, activity 3

Student B: You're Adam Hackett and you're a journalist. Here's some more information.

Mr Dave Dingle – student

Now turn back to page 11.

17 *Lesson 10*

Functions, activity 1

Student B: Look at the words for these things.

television   taxis   football

Now turn back to page 23.

18 *Lesson 11*

Functions and grammar, activity 3

Student B: Answer Student A's questions about 1-3. Then, ask Student A how much 4-6 are and write the price.

1 jeans — £20.00 4 sweater

2 jacket — £70.00 5 shirt

3 shoes — £34.99 6 skirt

Now turn back to page 27.

19 *Lesson 12*

Speaking and writing, activity 1

Student B: Answer Student A's questions.

Ask Student A where these objects are and put a cross [✗] on the picture.

coat   watch   glasses   book

Now turn back to page 29.

20 *Lesson 14*

Reading and speaking, activity 2

Student B: Match the questions with the missing information, then ask and answer to complete the E-mail. If you don't know how to spell the names, ask *How do you spell ...?*

How old is she?

What's he like?

What's her father's job?

What's her mother's name?

What's her brother's name?

(1) How old is she?

Dear Brad,

Thanks for your E-mail.

My name is Sue White. I'm an English teacher and I'm (1) ____ years old. I've got dark hair and brown eyes. My family is from London. My mother's name is (2)____ and she's a teacher. My father's name is Henry. He's a (3) _____. He's got (4) _____ eyes, but he hasn't got any hair! I've got a brother and a sister. My brother (5)____ is an engineer, too! He's twenty-two years old and he's got (6)____ hair (he's very (7)____)! My sister's name is Sarah. She's eleven; she's also got fair hair and blue eyes.

Write soon,

Sue

Now turn back to page 33.

21 *Lesson 17*

Listening and reading, activity 3

Student B: Listen and find out what time they have breakfast in Mexico, lunch in Russia and dinner in Hong Kong.

Now turn back to page 39.

22 *Lesson 5*

Writing and speaking, activity 2

Student C: Read the information.

Terry Crystal is American.

Maria Agnelli is from Naples.

The singer is Russian.

Now turn back to page 13.

23 *Lesson 19*

Vocabulary and sounds, activity 3

Student B: Look at the words for these sports.

Basketball   Gymnastics   Swimming   Baseball   Sailing

Now turn back to page 42.

24 *Lesson 22*

Vocabulary and sounds, activity 2

Student B: Look at the words for these things.

Now turn back to page 50.

25 *Lesson 22*

Grammar, activity 1

Student B: 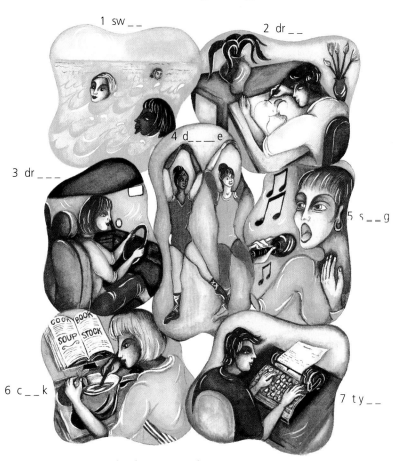 Listen and answer.

What do they drink in Morocco?

When is the main meal in India?

What does he have for breakfast?

Now turn back to page 51.

26 *Lesson 4*

Listening and speaking, activity 3

Student C: You're Miss Fifi Lamour and you're an actress. Here's some more information.

Mr Frank Fearless – actor

Now turn back to page 11.

27 *Lesson 17*

Listening and reading, activity 3

Student C: Listen and find out what time they have breakfast in Hong Kong, lunch in Mexico and dinner in Russia.

Now turn back to page 39.

28 *Lesson 24*

Speaking and writing, activity 1

Student A: Ask about Student B's living room.Use the chart below. Add other items of your own.

Is there a television in your living room?

Living room	Yes/No
armchair	
window	
table	
telephone	
sofa	
television	

Now answer Student B's questions.

Now turn back to page 55.

29 *Lesson 22*

Grammar, activity 1

Student C: Listen and answer.

When is the main meal in Morocco?

What typical food do they eat in India?

What does she have for breakfast?

Now turn back to page 51.

30 *Lesson 26*

Vocabulary and sounds, activity 2

Match the words with the drawings.

dance swim draw cook sing drive type

1 sw _ _

2 dr _ _

3 dr _ _ _

4 d _ _ _ e

5 s _ _ g

6 c _ _ k

7 t y _ _

Now turn back to page 58.

31 *Lesson 28*

Vocabulary and sounds, activity 3

Student A: Look at the pictures.

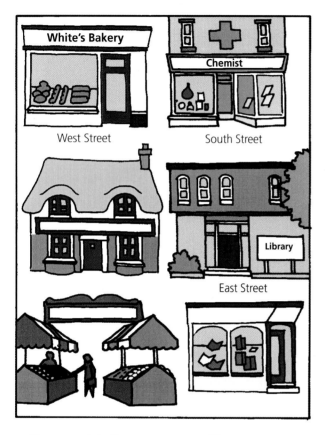

Now complete the map on page 62

32 *Lesson 29*

Speaking, activity 1

Think of the photos in Unit 29 and describe them in detail.

Now turn back to page 65.

33 *Lesson 30*

Speaking and writing, activity 1

Work together and say as many differences as you can between the pictures.

Now turn back to page 67.

34 *Lesson 33*

Speaking and writing, activity 1

Look at the photo for one minute only, then turn back to page 75.

35 *Lesson 24*

Speaking and writing, activity 1

Student B: Answer Student A's questions.

Now, ask about Student A's kitchen. Use the chart below. Add other items of your own.

Kitchen	Yes/No
cooker	
table	
chairs	
cupboards	
telephone	
television	

Is there a cooker in your kitchen?

Now turn back to page 55.

36 *Lesson 20*

Speaking, activity 1

Look at the card. Find out:

- who he/she is
- where he/she works
- what his/her job is
- where he/she lives

Merryfield School Library Card

Mary Ward
16 Green Street
Oxford

Now turn back to activity 2 on page 45.

37 *Lesson 38*

Reading and writing, activity 4

Dr Ueno was a teacher at a university in Tokyo. Every morning he walked with his dog, Hachiko, to the station to go to work. Hachiko walked home, but returned in the evening to meet Dr Ueno when he arrived at the station.

One day Hachiko arrived at the station and waited for Dr Ueno, but Dr Ueno didn't arrive that day. He died in the afternoon. Hachiko waited for six hours and then walked home.

The next day Hachiko returned to the station and waited for his friend Dr Ueno – and the next day and the next. He walked to the station every day at the same time – for nine years.

38 *Lesson 35*

Speaking and listening, activity 1

Look at the drawing from the nineteenth century.
What mistakes are there?

Now turn back to page 79.

39 *Lesson 39*

Writing and speaking, activity 2

Student B: Read the passage and answer the
questions you wrote in 1.

On Sunday morning, we left our hotel at (1) - in the
morning and took our (2) *cases* to the airport. Then we
went to the (3) - by boat. We spent (4) *an hour* there,
and then returned to Manhattan. We bought some (5) -
for lunch and ate them in (6) *Central Park.* We sat in the
park for about (7) -. In the afternoon, we went to the
(8) *cinema* and saw a (9) -. By now, it was time to go to
the airport for our flight home, so we took a (10) *taxi.*
We got the plane at (11) - and flew home to (12) *London.*
The end of a wonderful weekend!

Now turn back to page 87.

40 *Lesson 4*

Listening and speaking, activity 3

Student D: You're Tom James and you're a singer.
Here's some more information.

Mike Handy – engineer

Now turn back to page 11.

41 *Lesson 28*

Vocabulary and sounds, activity 3

Student B: Look at the pictures.

Pub

East Street

Market

Bookshop

North Street

West Street

Now complete the map on page 62.

42 *Lesson 40*

Reading and listening, activity 6

Then I saw something on the ground. I picked it up and looked at it. It was a railway ticket. It had a date on it: 28 October 1962.

Now turn back to activity 7 on page 89.

Grammar review

CONTENTS

Present simple

Form

You use the contracted form in spoken and informal written English.

Be

Affirmative	Negative
I'm (I am)	I'm not (am not)
you	you
we're (are)	we aren't (are not)
they	they
he	he
she's (is)	she isn't (is not)
it	it

Questions	Short answers
Am I?	Yes, I am.
	No, I'm not.
Are you/we/they?	Yes, you/we/they are.
	No, you/we/they're not.
Is he/she/it?	Yes, he/she/it is.
	No, he/she/it isn't.

Have

Affirmative	Negative
I	I
you've (have)	you haven't (have not)
we	we
they	they
he	he
she has	she hasn't (has not)
it	it

Questions	Short answers
Have I/you/we/they?	Yes, I/you/we/they have.
	No, I/you/we/they haven't.
Has he/she/it?	Yes, he/she/it has.
Has he/she/it?	No, he/she/it hasn't.

Regular verbs

Affirmative	Negative
I	I
you like	you don't (do not) like
we	we
they	they
he	he
she likes	she doesn't (does not) like
it	it

Questions	Short answers
Do I/you/we/they like?	Yes, I/you/we/they do.
	No, I/you/we/they don't (do not).
Does he/she/it like?	Yes, he/she/it does.
	No, he/she/it doesn't (does not).

Question words with *is/are*

What's your name?

Who's your favourite singer?

Question words with *does/do*

What do they eat in Morocco?

Where does he live?

Present simple: third person singular

You add -*s* to most verbs

leaves, starts

You add -*es* to *do, go* and verbs which end in *-ch, -ss, -sh* and *-x*

goes, does, watches, finishes

You add -*ies* to verbs ending in *-y*

carries, tries

Use

You use the present simple:

● to talk about customs. (See Lesson 17)

In Britain we have dinner at six o'clock in the evening.

In Thailand we have breakfast at seven o'clock in the morning.

● to talk about habits. (See Lesson 25)

I always get presents and birthday cards.

I usually go out with friends.

● to talk about routines. (See Lesson 20)

She leaves at 7.30 am and arrives at work at 8 am.

On Saturday afternoon, he goes shopping with their sons.

Present continuous

Form

You form the present continuous with *be* + present participle (-*ing*). You use the contracted form in spoken and informal written English.

Affirmative	Negative
I'm (am) walking	I'm not (am not) walking
you	you
we're (are) walking	we aren't (are not) walking
they	they
he	he
she's (is) walking	she isn't (is not) walking
it	it

Questions	Short answers
Am I walking?	Yes, I am.
	No, I'm not.
Are you/we/they walking?	Yes, you/we/they are.
	No, you/we/they aren't.
Is he/she/it walking?	Yes, he/she/it is.
	No, he/she/it isn't.

Question words

Where are they staying?   *Who is meeting them?*

Present participle (-*ing*) endings

You form the present participle of most verbs by adding -*ing*:

go – going, visit – visiting

You add -*ing* to verbs ending in *-e*:

make - making, have - having

You double the final consonant of verbs of one syllable ending in a vowel and a consonant, and add -*ing*:

get - getting, shop - shopping

You add -*ing* to verbs ending in a vowel and *-y* or *-w*:

draw - drawing, play - playing

You don't usually use these verbs in the continuous form.

believe feel hate hear know like love smell sound taste understand want

Use

You use the present continuous:

● to describe something that is happening at the moment. (see Lessons 29, 30)

He's buying lunch.

They're waiting for a bus.

- to talk about a definite arrangement in the future.
  (See Lesson 31)
  We're spending a week in a hotel.

Past simple

Form

You use the contracted form in spoken and informal written English.

Be

Affirmative		Negative	
I		I	
he	was	he	wasn't (was not)
she		she	
it		it	
you		you	
we	were	we	weren't (were not)
they		they	

Have

Affirmative		Negative	
I		I	
you		you	
we		we	
they	had	they	didn't (did not) have
he		he	
she		she	
it		it	

Regular verbs

Affirmative		Negative	
I		I	
you		you	
we		we	
they	listened	they	didn't (did not) listen
he		he	
she		she	
it		it	

Questions		Short answers		
	I		I	I
	you		you	you
	we		we	we
Did	they listen?	Yes,	they did. No,	they didn't.
	he		he	he
	she		she	she
	it		it	it

Question words

When did they arrive?
How much did Kate spend on her jacket?
How long did the journey take?
What did they buy?
How did they get to the hotel?

Past simple endings

You add *-ed* to most regular verbs:
walk - walked   watch - watched

You add *-d* to verbs ending in *-e*:
close - closed   continue - continued

You double the consonant and add *-ed* to verbs of one syllable ending in a vowel and a consonant:
stop - stopped   plan - planned

You drop the *-y* and and add *-ied* to verbs ending in *-y*:
study - studied   try - tried

You add *-ed* to verbs ending in a vowel and a *-y*
play - played

Irregular verbs

There are many verbs which have an irregular past simple. For a list of the irregular verbs which appear in **Reward Starter**, see page 108.

Pronunciation of past simple endings

/t/ *worked   finished*
/d/ *played   stayed*
/ɪd/ *wanted   visited*

Use

You use the past simple:

- to talk about a past action or event that is finished.
  (See Lessons 33 - 39)
  Picasso was born in Malaga.

- to talk about a state, habit or routine in the past
  (See Lessons 33 - 39)
  They went to New York for the weekend.
  We took a taxi from the airport.

Questions

You can form questions in two ways:

- without a question word
  Are you James Bond? (See Lesson 4)
  Is she married? (See Lesson 6)
  Does he go to work by boat? (See Lesson 21)
  Did you take a photograph? (See Lesson 38)

- with a question word *who, what, how, when, where*
  Who's your favourite singer? (See Lesson 8)
  What's your name? (See Lesson 1)
  How did they get to the hotel? (See Lesson 39)
  When did they arrive? (See Lesson 39)
  Where did they go? (See Lesson 39)

You can put an adjective or an adverb after *how*.
How much are they? (See Lesson 11)
How old is she? (See Lesson 7)
How long did the journey take? (See Lesson 39)

You can also form questions about ability with *can*.
(See Lesson 26)
Can you swim?
Can you use a computer?

You can form more indirect, polite questions with *can*.
Can I have a sandwich, please? (See Lesson 27)

Imperatives

The imperative has exactly the same form as the infinitive (without *to*) and does not usually have a subject. You use the imperative:

- to give instructions (See Lesson 15)
  Stand up!
  Don't talk.
  Open your book.

- to give directions (See Lesson 28)
  Go along South Street.
  Turn left.
  Turn right.

You use *don't* + imperative to give a negative instruction.
Don't talk!
Don't look!

Verb patterns

like + -ing form verb

You can put an *-ing* form verb after *like*. (See Lesson 23)
I don't like swimming.
My girlfriend likes sightseeing, but she doesn't like walking.
We don't like staying in hotels.

When you *like doing something*, this is something you enjoy all the time.
I like travelling by train. I always go by train.

Have got

You use the contracted form in spoken and informal written English.

Affirmative	Negative
I	I
you 've (have) got	you haven't (have not) got
we	we
they	they
he	he
she's (has) got	she hasn't (has not) got
it	it

Questions	Short answers
Have I/you/we/they got?	Yes, I/you/we/they have.
	No, I/you/we/they haven't.
Has he/she/it got?	Yes, he/she/it has.
	No, he/she/it hasn't.

You use *have got*

- to talk about possession (See Lesson 13)
  We've got three children.

- to talk about appearance (See Lesson 14)
  She's got fair hair and blue eyes.
  He hasn't got any hair.

  Have got means the same as *have*. You use it in spoken and informal written English.
  She's got fair hair and blue eyes. (= She has fair hair and blue eyes)

Modal verbs

Can and *need* are modal verbs. Other modal verbs are *could must should will would*.

Form

Modal verbs:

- have the same form for all persons.
  I can swim.
  He can play the piano.

- don't take the auxiliary *do* in questions and negatives.
  Can you cook?
  I can't drive.

- take an infinitive without *to*.
  He can dance.
  She can cook.

Use

You use *can:*

- to talk about general ability, something you are able to do on most occasions. (See Lesson 26)
  I can swim.   I can play the piano.

- to ask for something politely. (See Lesson 27)
  Can I have a sandwich, please?
  Can I have a cup of coffee?

Articles

There are many rules for the use of articles. Here are the rules presented in **Reward Starter.**

You use an indefinite article (*a/an*) with jobs. (See Lesson 2)
I'm a student.
I'm a doctor.
I'm an actor.
I'm an engineer.

You use *an* for nouns which begin with a vowel.
an actor   an engineer

You use the definite article (*the*):

- when the listener or reader knows exactly which person or thing we mean (See Lesson 11)
  How much are the black shoes?
  How much is the green sweater?

You also use *the* when you talk in general about musical instruments.
He can play the piano.
She plays the guitar.
Before vowels you pronounce *the /ə/.*

Plurals

You form the plural of most nouns with -*s* (See Lesson 9)
Singular: *friend   neighbour   boy   girl   brother   sister   twin*
Plural: *friends   neighbours   boys   girls   brothers   sisters   twins*

You add -*es* to nouns which end in -*o*, -*ch*, -*ss*, -*sh* and -*x*
glass - glasses   watch - watches   sandwich - sandwiches

There are some irregular plurals:
man - men   woman - women   child - children

Expressions of quantity

Any

You use *any* with *there aren't* and *are there?* (See Lesson 24)
There aren't any cupboards in the bathroom.
Are there any chairs in the kitchen?

Possessives

Possessive -s

You add -*s* to singular nouns to show possession.
 (See Lesson 12)
Jane's keys   Graham's wallet
You add -*s'* to regular plural nouns.
My brothers' names are Pablo and Octavio.
You add -*'s* to irregular plural nouns.
The men's room.

Possessive adjectives

Pronoun	Possessive adjective	
I	my	(See Lesson 1)
you	your	(See Lesson 1)
he	his	(See Lesson 8)
she	her	(See Lesson 8)
it	its	
we	our	(See Lesson 13)
they	their	(See Lesson 13)

Adjectives

Position of adjectives

You can put adjectives in two positions.

- after the verb *to be* (See Lesson 5 and 33)
  He's British. She's Brazilian.
  It was warm. He was happy.

- before a noun (See Lesson 11)
  the black jacket   the blue jeans   the green sweater

Demonstrative adjectives

This, that, these and those (See Lesson 10)

You use *this* to point to singular nouns which are close.
What's this?
It's a watch.

You use *that* to point to singular nouns which are not close.
What's that?
It's an umbrella.

You use *these* to point to plural nouns which are close.
What are these?
They're glasses.

You use *those* to point to plural nouns which are not close.
What are those?
They're books.

Adverbs

Position of adverbs of frequency

You usually put adverbs of frequency before the verb. (See Lesson 25)

I always get presents and birthday cards.
I usually go out with friends.
I often go to a restaurant.

But you put them after the verb *to be*.

I never do anything special..
She was always cold in December.

Prepositions

from, in, to, at, on

You use *from:*

● with towns and countries to talk about people's homes. (See Lesson 5)
He's from London.
I'm from Bangkok.

You use *in:*

● to describe position. (See Lesson 12)
Graham's wallet is in his coat pocket.

● with places (See Lesson 16)
I live in a flat in Florence.

● with times of the day
in the morning, in the afternoon, in the evening

You use *to:*

● with *go + school and work*. (See Lesson 16)
I go to school in Fiesole.

You use *at*:

● with times of the day. (See Lesson 17)
We have lunch at two o'clock.

You use *on*:

● to describe position. (See Lesson 12)
Steve's wallet is on the table.

● with days of the week (See Lesson 18)
on Monday   on Tuesday   on Monday morning
on Tuesday afternoon

You use *under*:

● to describe position. (See Lesson 12)
Joely's bag is under the table.

You use *by*:

● with means of transport (See Lesson 21)
by car   by train   by bus   by taxi

Pronunciation guide

/ɑː/	p<u>ar</u>k	/b/	<u>b</u>uy
/æ/	h<u>a</u>t	/d/	<u>d</u>ay
/aɪ/	m<u>y</u>	/f/	<u>f</u>ree
/aʊ/	h<u>ow</u>	/g/	<u>g</u>ive
/e/	t<u>e</u>n	/h/	<u>h</u>ouse
/eɪ/	b<u>ay</u>	/j/	<u>y</u>ou
/eə/	th<u>ere</u>	/k/	<u>c</u>at
/ɪ/	s<u>i</u>t	/l/	<u>l</u>ook
/iː/	m<u>e</u>	/m/	<u>m</u>ean
/ɪə/	b<u>eer</u>	/n/	<u>n</u>ice
/ɒ/	wh<u>a</u>t	/p/	<u>p</u>aper
/əʊ/	n<u>o</u>	/r/	<u>r</u>ain
/ɔː/	m<u>ore</u>	/s/	<u>s</u>ad
/ɔɪ/	t<u>oy</u>	/t/	<u>t</u>ime
/ʊ/	t<u>oo</u>k	/v/	<u>v</u>erb
/uː/	s<u>oo</u>n	/w/	<u>w</u>ine
/ʊə/	t<u>our</u>	/z/	<u>z</u>oo
/ɜː/	s<u>ir</u>	/ʃ/	<u>sh</u>irt
/ʌ/	s<u>u</u>n	/ʒ/	lei<u>s</u>ure
/ə/	bett<u>er</u>	/ŋ/	si<u>ng</u>
		/tʃ/	<u>ch</u>ur<u>ch</u>
		/θ/	<u>th</u>ank
		/ð/	<u>th</u>en
		/dʒ/	<u>j</u>acket

Irregular Verbs

Verbs with the same infinitive, past simple and past participle

cost	cost	cost
cut	cut	cut
hit	hit	hit
let	let	let
put	put	put
read /ri:d/	read /red/	read /red/
set	set	set
shut	shut	shut

Verbs with the same past simple and past participle but a different infinitive

bring	brought	brought
build	built	built
burn	burnt/burned	burnt/burned
buy	bought	bought
catch	caught	caught
feel	felt	felt
find	found	found
get	got	got
have	had	had
hear	heard	heard
hold	held	held
keep	kept	kept
learn	learnt/learned	learnt/learned
leave	left	left
lend	lent	lent
light	lit/lighted	lit/lighted
lose	lost	lost
make	made	made
mean	meant	meant
meet	met	met
pay	paid	paid
say	said	said
sell	sold	sold
send	sent	sent
sit	sat	sat
sleep	slept	slept
smell	smelt/smelled	smelt/smelled
spell	spelt/spelled	spelt/spelled
spend	spent	spent
stand	stood	stood
teach	taught	taught
understand	understood	understood
win	won	won

Verbs with same infinitive and past participle but a different past simple

become	became	become
come	came	come
run	ran	run

Verbs with a different infinitive, past simple and past participle

be	was/were	been
begin	began	begun
break	broke	broken
choose	chose	chosen
do	did	done
draw	drew	drawn
drink	drank	drunk
drive	drove	driven
eat	ate	eaten
fall	fell	fallen
fly	flew	flown
forget	forgot	forgotten
give	gave	given
go	went	gone
grow	grew	grown
know	knew	known
lie	lay	lain
ring	rang	rung
rise	rose	risen
see	saw	seen
show	showed	shown
sing	sang	sung
speak	spoke	spoken
swim	swam	swum
take	took	taken
throw	threw	thrown
wake	woke	woken
wear	wore	worn
write	wrote	written

Tapescripts

Lesson 1 **Listening and reading, activity 2**

ANNA Hello, I'm Anna. What's your name?
DAVID Hello, Anna. I'm David.
TONY Hello, I'm Tony. What's your name?
JANE Hello, Tony. I'm Jane.
JUDY Hello, I'm Judy. What's your name?
STEVE Hello, Judy. I'm Steve.

Lesson 2 **Vocabulary and sounds, activity 2**

JOSÉ What's your job, Pete?
PETE I'm a journalist. What's your job, José?
JOSÉ I'm a doctor.
MARIA What's your job, Hillary?
HILLARY I'm a secretary. What's your job, Maria?
MARIA I'm a teacher.
HASHIMI What's your job, Yıldız?
YILDIZ I'm a student. What's your job, Hashimi?
HASHIMI I'm an engineer.

Lesson 2 **Listening, activity 2**

MARIA What's your job, Hillary?
HILLARY I'm a secretary. What's your job, Maria?
MARIA I'm a teacher.

Lesson 3 **Listening and reading, activity 2**

Conversation A
MICHIKO Hello, Joan, how are you?
JOAN Hello, Michiko, I'm very well, thanks. How are you?
MICHIKO I'm fine, thanks.

Conversation B
PETE Kate, what's your telephone number, please?
KATE 0134 521 3987. What's your telephone number, Pete?
PETE 01967 328123.
KATE Thank you. Goodbye, Pete.
PETE Goodbye.

Lesson 4 **Listening and speaking, activity 1**

r-a-m-b-o
c-l-e-o-p-a-t-r-a
d-r-a-c-u-l-a
f-r-a-n-k-e-n-s-t-e-i-n
j-a-m-e-s- b-o-n-d

Lesson 5 **Vocabulary and sounds, activity 5**

American - The USA
Japanese - Japan
Italian - Italy
Brazilian - Brazil
Thai - Thailand
British - Britain
Turkish - Turkey
Russian - Russia.

Lesson 5 **Listening and reading, activity 3**

OLGA Hello, I'm Olga Maintz. I'm an engineer. I'm Russian and I'm from St Petersburg.
MUSTAFA Hi! I'm Mustafa Polat. I'm Turkish and I'm a teacher. I'm from Istanbul.
PATRIZIO Hello! I'm Patrizio Giuliani. I'm from Venice. I'm Italian and I'm an actor.

Lesson 6 **Vocabulary and sounds, activity 3**

One Two Thirteen Fourteen Five Six Seventeen Eighteen Nine Twenty

Lesson 6 **Listening and writing, activity 2**

Is Ken Stanwell from Kenton?
Yes, he is.
Is he married?
No, he isn't. He's seventeen.
Is he a student?
Yes, he is.
And is he British?
No, he isn't. He's American.

Lesson 6 **Grammar, activity 3**

Is Jane married?
Yes, she is.
Is Anna married?
No, she isn't.
Is Sema married?
No, she isn't.
Is Kazuo married?
Yes, he is.
Is Steve married?
Yes, he is.

Lesson 6 **Speaking and listening, activity 2**

Is Bill Clinton an engineer?
No, he isn't.
Is Tom Cruise an actor ?
Yes, he is.
Is pizza from Italy?
Yes, it is.
Is doctor a job?
Yes, it is.
Is San Francisco in the United States?
Yes, it is.
Are you from Japan?
No, I'm not./ Yes, I am.
Is Graham a French name?
No, it isn't.
Is Istanbul a country?
No, it isn't.
Is seventeen one-seven?
Yes, it is.
Is Whitney Houston American?
Yes, she is.
Are you President of the USA?
No, I'm not.
Is Argentinian a nationality?
Yes, it is.
Is Edinburgh in England?
No, it isn't.
Is Spain a country?
Yes, it is.
Is Roberto Baggio an actor?
No, he isn't.
Are you André Agassi?
No, I'm not.

Is Sony Korean?
No, it isn't.
Is champagne from France?
Yes, it is.
Is your name Queen Elizabeth?
No, it isn't.
Are you married.
Yes, I am. Are you married?
No, I'm not.

Lesson 7   Vocabulary and sounds, activity 3

Thirteen   Forty   Fifty   Sixteen   Seventeen   Eighty   Nineteen

Lesson 7   Listening, activity 1

How old is Tony?
He's twenty.
How old is Karen?
She's twenty-seven.
How old is Nick?
He's nineteen.
How old is Sarah?
She's twenty-three.
How old is Jill?
She's thirty-five.
How old is Alex?
He's seventeen.

Lesson 7   Listening and writing, activity 1

Ella isn't married. She's 42. She's an actress and she's American.
Miki's a student. She's 18. She isn't married and she's Japanese.
Maria's 24. She's Italian and she's a journalist. She isn't married.
Erol's Turkish. He's married and he's a waiter. He's 21.
Anant's married and he's a teacher. He's 30.
Carlos is 33. He's Brazilian and he's a doctor. He isn't married.

Lesson 8   Vocabulary and sounds, activity 2

One.   Car
Two.   Football team
Three.   TV programme
Four.   TV presenter
Five.   Group
Six.   Politician

Lesson 8   Writing and listening, activity 2

SALLY What's your favourite car, Max?
MAX My favourite car? A Porsche, I think. Yes, a Porsche.
SALLY And what's your favourite football team?
MAX Manchester .
SALLY Manchester United or Manchester City?
MAX Machester United, of course. What about you? Who's your favourite actor?
SALLY It's Arnold Schwarzenegger.
MAX Arnold Schwarzenegger, I see. And what's your favourite group?
SALLY Well, my favourite singer is Diana Ross. What's your favourite group?
MAX The Beatles.
SALLY The Beatles! Who are the Beatles?

Lesson 9   Reading and listening, activity 20

JANE Are you twins?
NICK Yes, we are. I'm Nick.
DAVE And I'm Dave.
JANE Are you from London?
NICK/DAVE No, we aren't. We're from Manchester.
JANE How old are you?

NICK/DAVE We're twenty-three.
JANE What are your jobs?
NICK/DAVE We're students.
JANE What's your favourite football team?
NICK/DAVE Manchester United!

Lesson 9   Reading and listening, activity 3

PAUL Who are they?
JANE They're ... er Nick and Dave.
PAUL Are they from London?
JANE No, they're from, er, ... Manchester.
PAUL And how old are they?
JANE They're twenty-three. They're students.

Lesson 10   Vocabulary and sounds, activities 2 and 3

One.   They're cassettes.
Two.   They're glasses.
Three.   It's a wallet.
Four.   It's a pen.
Five.   It's a clock.
Six.   It's an umbrella.
Seven.   They're keys.
Eight.   They're books.
Nine.   It's a watch.
Ten.   It's a bag.

Progress check, lessons 1 to 10

Reading and listening, activity 2

One, two, three o'clock, four o'clock rock,
Five, six, seven o'clock, eight o'clock rock,
Nine, ten, eleven o'clock, twelve o'clock rock,
We're gonna rock around the clock tonight.
Put your glad rags on and join me, hon'
We'll have some fun when the clock strikes one,
Chorus
We're gonna rock around the clock tonight,
We're gonna rock, rock, rock til broad daylight
We're gonna rock, gonna rock around the clock tonight.
When the clock strikes two, three and four,
If the band slows down, we'll yell for more
Chorus
When the chimes ring five and six and seven
We'll be rockin' up in seventh heaven.
Chorus
When it's eight, nine, ten, eleven, too
I'll be going strong and so will you,
Chorus
When the clock strikes twelve, we'll cool off, then,
Start a rocking round the clock again.
Chorus

Lesson 11   Vocabulary, activity 4

MAN 1 How much is the jacket?
WOMAN 1 Fifty pounds.
MAN 2 How much are the jeans?
WOMAN 1 Twenty-seven, ninety-nine.
WOMAN 2 How much is the skirt?
WOMAN 1 Thirty-five, ninety-nine.
WOMAN 2 How much is the sweater?
WOMAN 1 Twenty one, fifty.
MAN 3 How much is the shirt?
WOMAN 1 Twelve, fifty.
WOMAN 2 How much are the shoes?
WOMAN 1 Forty, ninety-nine.

Lesson 11   Listening and sounds, activities 1 and 2

CUSTOMER How much are these black shoes?
ASSISTANT They're £29.50.
CUSTOMER And how much is that red sweater?
ASSISTANT It's £17.
CUSTOMER How much is this blue jacket?
ASSISTANT It's £40.
CUSTOMER How much are these black jeans?
ASSISTANT They're £35.99.

Lesson 11   Listening and speaking, activity 2

MAN How much is a big Mac in Britain?
WOMAN It's about two pounds.
MAN And how much are Levi jeans?
WOMAN They're thirty pounds.
MAN And Nike trainers? How much are they?
WOMAN They're forty pounds.
MAN And how much is a Cola?
WOMAN About thirty-five pence.

Lesson 12   Listening, activities 1 and 3

MAN Where're Jane's keys?
WOMAN They're on the chair.
MAN Where's Graham's wallet?
WOMAN It's in his coat pocket.
MAN And where's Frank's watch?
WOMAN It's on the chair.
MAN And where's Joely's bag?
WOMAN It's under the table.
MAN And where are Nicola's glasses?
WOMAN They're on the table.
MAN And where's Tom's personal stereo?
WOMAN It's on the table.

Lesson 12   Grammar, activity 3

TOM Where's my personal stereo?
JOELY It's on the table.
JOELY Where's my bag?
NICOLA It's under the table.
GRAHAM Where's my wallet?
JANE It's in your coat pocket.

Lesson 13   Listening and speaking, activity 2

INTERVIEWER Are you maried, Marco?
MARCO Yes, I am.
INTERVIEWER Have you got any children?
MARCO No, I haven't.
INTERVIEWER Have you got any brothers or sisters?
MARCO Yes, I have. I've got two brothers and one sister.
INTERVIEWER What are their names?
MARCO Paolo, Giovanni, and Patrizia.

Lesson 15   Vocabulary and listening, activity 6

Come in.
Take your coat off.
Sit down.
Open your book.
Pick your pen up.
Turn the cassette player on.

Lesson 15   Sounds and speaking, activity 3

Stand up.
Open your book.
Pick your pen up.
Sit down.

Close your book.
Put your pen down.
Pick your bag up.
Put your coat on.
Put your bag down.
Take your coat off.
Sit down.

Lesson 16   Reading and listening, activity 3. Photo 1

BOY Hello, I'm Erol. I'm from Turkey and I'm sixteen. I'm a student. My sister's name is Belma and we go to school in Istanbul. We live with our parents in a flat in Galata. My father is an engineer and my mother is a secretary. They work in an office in Beyoğlu.
Photo 2
MAN Hello, I'm Kazuo. I'm from Japan and I'm thirty-five. I'm a journalist and I work in an office in Tokyo. My wife's name is Michiko and we live in a flat in Ichikawa. We've got two children. Our son's name is Koji and our daughter's name is Miki. They go to school in Funabashi.

Lesson 17   Speaking and vocabulary, activity 4

MAN Hello, everybody and welcome to the Afternoon show. It's three o'clock. We've got some very interesting guests for you today ...
WOMAN This is the BBC World Service. It's eleven o'clock and here is the news.
MAN It's six o'clock and here is the news.
WOMAN And that was the end of the one clock news. Here's the weather forecast for all areas for the next twenty-four hours ...

Lesson 17   Listening and reading, activity 3

INTERVIEWER When do you have your meals in Russia?
MAN We have breakfast at seven o'clock, lunch at twelve o'clock and dinner at six o'clock.
INTERVIEWER And what about in Hong Kong? When do you have breakfast?
WOMAN 1 Oh, about seven o'clock in the morning. And we have lunch at about one o'clock.
INTERVIEWER And when do you have dinner?
WOMAN 1 About eight o'clock.
INTERVIEWER And when do you have breakfast in Mexico?
WOMAN 2 We have breakfast at about eight o'clock. Then we have lunch at two or three o'clock in the afternoon.
INTERVIEWER And dinner?
WOMAN 2 At about eight or nine o'clock in the evening.

Lesson 19   Listening and writing, activity 3

TIM Do you like gymnastics, Gwen?
GWEN Yes, I do. I like gymnastics very much.
TIM Do you like basketball?
GWEN No, I don't.

JAMES Do you like volleyball, Alison?
ALISON Yes, I do. I like volleyball very much.
JAMES Do you like skiing?
ALISON No, I don't.

Lesson 20   Vocabulary and speaking, activity 4

MAN It's half past five. Time to go home.
DJ Good morning! Good morning! It's a quarter to seven and this is the Pete Walker show...
TV PRESENTER It's a quarter past twelve. In fifteen minutes time we'll be talking to ...
MUM It's a quarter to six. Hurry up! We're leaving in five minutes.
DAD Get up, John! It's half past seven.
WOMAN Time for bed. It's a quarter past eleven. Goodnight!

Lesson 20 **Listening and writing, activity 1**

MAN1 Hello, and welcome to The British Abroad, the programme where we talk to British people living and working in other countries. Where do you live, Sarah?
WOMAN I live in Italy.
MAN 1 And where do you work?
WOMAN I work in an English school.
MAN 1 When do you have breakfast in Italy?
WOMAN At about half past seven.
MAN 1 When do you start work?
WOMAN I start work at three in the afternoon.
MAN 1 When do you finish work?
WOMAN At nine o'clock in the evening.
MAN 1 When do you go shopping in Italy?
WOMAN In the mornings and on Saturdays.
MAN 1 When do you visit friends in Italy?
WOMAN On Saturday evenings and on Sundays.
MAN 1 Where do you live, Mark?
MAN 2 I live in Acapulco.
MAN 1 And where do you work?
MAN 2 I work in a hospital. I'm a doctor.
MAN 1 When do you have breakfast?
MAN 2 At about half past six.
MAN 1 And when do you start work?
MAN 2 I start work at eight in the morning.
MAN 1 When do you finish work?
MAN 2 At seven o'clock in the evening.
MAN 1 When do you go shopping?
MAN 2 My wife goes shopping on Saturdays.
MAN 1 When do you visit your parents?
MAN 2 My parents live in Britain but my wife's parents live in Acapulco, too. We visit them on Sundays.

Progress check, lessons 11 to 20. **Listening, activities 1 and 2**

Don't worry about a thing
'Cause every little thing's gonna be all right
Singing 'Don't worry about a thing
'Cause every little thing's gonna be all right.'
Rise up this morning
Smiled with the rising sun
Three little birds beside my doorstep
Singing sweet songs of melodies pure and true
Singing 'This is my message to you.'

Lesson 21 **Reading and listening, activity 2**

SALLY Thank you Mehmet ... Excuse me!
LEYLA Yes.
SALLY Your name is...?
LEYLA My name is Leyla.
SALLY Do you walk to work, Leyla?
LEYLA Yes, I do.
SALLY And are you married?
LEYLA No, I'm not.
SALLY Have you got any brothers or sisters?
LEYLA Yes, I have a brother. His name's Mustafa.
SALLY And does he walk to work?
LEYLA No, he goes to work by boat.
SALLY Thank you very much. Excuse me, sir! What's your name?
JOHN My name's John.
SALLY Ah, you're American. Do you live here?
JOHN Yes, I do.
SALLY And do you work here?
JOHN No, I don't. I work in Asia.
SALLY And do you go to work by car?
JOHN No, I go by boat. My wife, Mary, goes to work by car.
SALLY Thank you, John. So, as you can see, people here are using many different ways of going to work...

Lesson 22 **Listening, activity 1**

RADIO PRESENTER Hello and welcome to food and drink around the world. Today we're looking at food in Argentina, Morocco, and India. Christina is from Argentina, what food do you eat in your country, Christina?
CRISTINA In Argentina, we eat meat, especially beef.
RADIO PRESENTER What do you drink?
CRISTINA We drink beer and wine with our meals, or water and juice.
RADIO PRESENTER When do you have the main meal of the day?
CRISTINA The main meal of the day is lunch. We have lunch at one o'clock in the afternoon. During the week, if you work, the main meal is dinner.
RADIO PRESENTER What do you have for breakfast?
CRISTINA We have coffee and bread.

Lesson 22 **Grammar, activity 1**

RADIO PRESENTER Nourredine, tell me what you eat in Morocco.
NOURREDINE Well, a typical dish is a dish with lamb and vegetables.
RADIO PRESENTER And what do you drink?
NOURREDINE Our national drink is tea and we drink a lot of juice as well.
RADIO PRESENTER And when do you have the main meal of the day?
NOURREDINE The main meal in Morocco is lunch. We have it at twelve o'clock.
RADIO PRESENTER And what do you have for breakfast?
NOURREDINE We have milk and yoghurt, and fruit: apples and oranges.
RADIO PRESENTER Ram, what do you eat in India?
RAM We eat alot of rice and vegetables.
RADIO PRESENTER And what do you drink?
RAM Tea or water.
RADIO PRESENTER And when do you have the main meal of the day?
RAM At 10 am in the morning or 7 pm in the evening.
RADIO PRESENTER And what do you have for breakfast?
RAM Tea.

Lesson 23 **Reading and listening, activities 3 and 4**

INTERVIEWER Gary, what do you like doing on holiday?
GARY Well, I like walking and sightseeing.
INTERVIEWER Anything else?
GARY Er, I like eating in restaurants and staying in hotels.
INTERVIEWER Is there anything you don't like doing?
GARY I don't like skiing. I don't like lying on the beach or swimming. Oh, and I don't like writing postcards.
INTERVIEWER Margaret, what do you like doing on holiday?
MARGARET I like lying on the beach and reading. I like swimming as well. And in the evening I like dancing.
INTERVIEWER And what don't you like doing?
MARGARET I don't like sightseeing.
INTERVIEWER Do you like writing postcards?
MARGARET No, I don't.

Lesson 24 **Vocabulary and sounds, activity 6**

Kitchen - cooker, table, cupboards, chair.
Dining room - table, chairs.
Living room - armchairs, sofa, television, table.
Bedroom 1 - armchair, bed.
Bedroom 2- bed, table, cupboard.
Bathroom - shower.
Hall - table, telephone.

Lesson 24 **Reading and listening, activity 2**

George Mandelson is a journalist. He isn't married. He lives in a small house. He doesn't have a garden. He likes watching football on TV, dancing and taking photos.
Angie Ashton is a singer. She's married with four children. She likes cooking, seeing her friends, playing tennis and reading newspapers. She

doesn't eat meat and she doesn't like watching television.
Frances Peters is an actress. She's married but she hasn't got any children.
She lives in London. She likes seeing her friends, but she doesn't like cooking.

Lesson 25 Vocabulary and sounds, activity 2

January February March April May June July August September October November December.

Lesson 25 Vocabulary and sounds, activity 6

The seventh of July.
The eighth of August.
The ninth of September.
The tenth of October.
The eleventh of November.
The twelfth of December.

Lesson 25 Listening and speaking, activity 2

INTERVIEWER What do you for your birthday, Karen?
KAREN I usually have a party and I always get presents and birthday cards.
INTERVIEWER And Pete, what do you do?
PETE I usually go out with friends or I sometimes invite friends home.
INTERVIEWER Do you always get presents and birthday cards?
PETE Yes, I do.
INTERVIEWER What do you do for your birthday, Molly?
MOLLY I usually have a meal with my family. I often go to a restaurant and I always get presents and birthday cards.

Lesson 26 Reading and listening, activity 3

INTERVIEWER So, are you a student?
FRANK Yes, I am.
INTERVIEWER And you need a holiday job?
FRANK Yes, I do.
INTERVIEWER Can you swim?
FRANK Yes, I can.
INTERVIEWER And music? Can you play the piano?
FRANK No, I can't. But I can play the guitar.
INTERVIEWER OK, and can you speak any languages?
FRANK Yes, I can. I can speak French and German.
INTERVIEWER And what about sport. Can you play tennis and football?
FRANK Yes, I can.
INTERVIEWER Good.

Lesson 26 Listening and speaking, activity 1. Interview 1

INTERVIEWER Yes, come in.
JANIE Good morning.
INTERVIEWER Good morning. Come in and sit down. Take off your coat. What's your name?
JANIE Janie Ellis.
INTERVIEWER Are you a student?
JANIE Yes, I am. I need a holiday job at the moment.
INTERVIEWER Can you swim?
JANIE Yes, I can.
INTERVIEWER Can you play the piano?
JANIE Yes, I can.
INTERVIEWER Can you speak any languages?
JANIE I can speak a little Spanish.
INTERVIEWER And can you play tennis?
JANIE Yes, I can.
INTERVIEWER And what about football?
JANIE No, I can't.
INTERVIEWER All right. Where do you live, Janie? (FADE)

Interview two.

INTERVIEWER And your name is ...?
LOIS Lois Franks.
INTERVIEWER And you're a student?

LOIS Yes, I am. I'm a music student.
INTERVIEWER A music student! Excellent. So you can play the piano?
LOIS Yes, I can. I can play the piano, the guitar, the violin, the trumpet . . .
INTERVIEWER Excellent. And can you swim?
LOIS Yes, I can.
INTERVIEWER And can you speak any languages?
LOIS Yes, I can speak French, Italian, Russian and Spanish.
INTERVIEWER And do you like football and tennis?
LOIS Yes, they're my favourite sports.
INTERVIEWER Very good, Lois. Now, where are you from?

Lesson 27 Listening, activity 2

WAITER Can I help you?
JANE Yes, what's a Halley Court Special?
WAITER It's a sandwich with chicken, lettuce, tomato and mayonnaise.
JANE How much is it?
WAITER It's £2.50.
JANE OK, can I have a Halley Court Special, please?
WAITER Certainly. And anything to drink?
JANE A cup of coffee, please.
WAITER OK, a Halley Court Special and a cup of coffee. Anything else?
JANE A piece of chocolate cake, please.
WAITER Thank you . . . OK, a Halley Court Special, a cup of coffee and a piece of chocolate cake. Here you are.
JANE Thank you very much.
WAITER Enjoy your meal.

Lesson 27 Listening and speaking, activity 2

FINN What's your favourite food, Selina?
SELINA Well, I don't eat meat, so I like lots of salad things: lettuce and tomatoes etc, and vegetables. I like pizza and pasta very much,too, and I eat a lot of cheese. I like potatoes, but I don't eat them often.
FINN And what's your favourite drink?
SELINA I like mineral water and Cola. I don't like coffee, but I like tea. What about you? What's your favourite food?
FINN Well, I like chicken but I don't like beef. I like tomatoes and lettuce as well. I like potatoes, especially baked potatoes with tuna and mayonnaise.
SELINA What do you like to drink?
FINN Oh, I like tea. Do you know what my favourite food is, though?
SELINA No, what?
FINN Chocolate cake!
SELINA Oh yes! I like chocolate cake, too.

Lesson 27 Listening and speaking, activity 3

SELINA Can I have a cheese and tomato pizza and a mineral water, please?
WAITER Certainly. And for you, sir?
FINN Can I have a baked potato with tuna and mayonnaise?
WAITER Certainly. Anything to drink, sir?
FINN Oh, a cup of tea, please.
WAITER OK. That's a cheese and tomato pizza and a mineral water, one baked potato with tuna and mayonnaise, and a cup of tea. Anything else?
FINN Selina, look at that chocolate cake!
SELINA Wow! Great!
FINN And two pieces of chocolate cake, please.
WAITER Thank you.

Lesson 28 Listening, activity 1

MAN 1 Where's the station?
WOMAN 1 It's in East Street.
WOMAN 2 Where's the bookshop?
MAN 2 It's in West Street.
MAN 3 Where's the market?
WOMAN 3 It's in North Street.
WOMAN 4 Where's the chemist?
MAN 4 It's in South Street.

Lesson 28 **Listening, activity 2**

MAN 1 Where's North Street?
WOMAN 1 Go along West Street. Turn right into North Street.
WOMAN 2 Where's West Street?
MAN 2 Go straight ahead.
WOMAN 3 Where's South Street?
MAN 3 Turn right.
MAN 4 Where's East Street?
WOMAN 4 Turn right into South Street. Turn left into East Street.

Lesson 28 **Listening, activity 3**

MAN 1 Excuse me! Where's the post office?
WOMAN 1 It's in West Street.
MAN 1 Where's West Street?
WOMAN 1 Go straight ahead. It's on the left.
WOMAN 2 Where's the bank?
MAN 2 It's in East Street.
WOMAN 2 Where's East Street?
MAN 2 Turn right into South Street, and turn left into East Street. It's on the right.
MAN 3 Excuse me! Where's the cinema?
WOMAN 3 It's in South Street. Turn left and it's on the left.
MAN 3 Thank you.
WOMAN 4 Where's the car park?
MAN 4 It's in East Street. Go straight ahead, along West Street and turn right into North Street. Go along North Street and turn right into East Street. It's on the left.
WOMAN 4 Thank you.

Lesson 28 **Listening and reading, activity 1**

TOUR GUIDE Good morning, ladies and gentlemen, and welcome to Reward Walking Tours of London. I'm Jamie and I'm your guide this morning for our tour of Covent Garden and Trafalgar Square. Well, as you know, we're in Trafalgar Square, and the building over there is the National Gallery, and the building on my right is South Africa House. And the church between the National Gallery and South Africa House is St Martins-in-the-Field. Now, the street on your right here is Whitehall, with Big Ben and the Houses of Parliament in the distance, and on your left is the Mall, with Buckingham Palace at the end.

Lesson 29 **Sounds, activity 1**

stand in, reading, shopping in, sit in, playing, running, lie in, writing

Lesson 30 **Listening and vocabulary, activity 3**

Picture 1	What's he doing?
	Is he listening to the radio?
Picture 2	What's she doing?
	Is she making tea?
Picture 3	What are they doing?
	Are they waiting for a bus?
Picture 4	What's he doing?
	Is he having a bath?
Picture 5	What's she doing?
	Is she talking to her daughter?
Picture 6	What are they doing?
	Are they having dinner?

Lesson 30 **Listening and vocabulary, activity 6**

one	He's listening to the cd.
two	She's making coffee.
three	They're waiting for a taxi.
four	He's having a shower.
five	She's talking to her son.
six	They're having lunch.

Progress check, lessons 21 to 30 **Sounds, activity 2**

bicycle vegetable telephone restaurant library
America December September computer

Progress check, lessons 21 to 30 **Sounds, activity 3**

Asia February November sightseeing eleventh

Progress check, lessons 21 to 30 **Sounds, activity 5**

MAN They aren't playing football. They're playing tennis.
WOMAN He isn't eating. He's drinking.
MAN She isn't washing. She's cooking.

Progress check, lessons 21 to 30

Listening, activities 1 and 2

Daniel is travelling tonight on a plane
I can see the red tail lights heading for Spain,
Oh, and I can see Daniel waving goodbye,
God, it looks like Daniel, must be the clouds in my eyes.
They say Spain's pretty, though I've never been,
Well, Daniel says it's the best place that he's ever seen,
Oh and he should know he's been there enough,
Lord, I miss Daniel, oh I miss him so much.
Oh, Daniel my brother,
You are older than me,
Do you still feel the pain
Of the scars that won't heal?
Your eyes have died, but you see more than I,
Daniel, you're a star in the face of the sky.

Lesson 31 **Reading and listening, activity 2**

TONY What are your plans for the holiday? Where are you going?
DON We're going to Australia.
TONY Australia!
SUE Yes, for three weeks. We're staying with Fran in Sydney. We're spending a week there.
DON And then we're driving to Port Stephens, about three hours away, and we're spending a week on the beach.
TONY Wonderful!
SUE Then we're going to a town called Cobar in the outback. We're camping there.
TONY Well, have a great time!
DON Thanks. What are you doing?
TONY Oh, I'm staying at home.

Lesson 31 **Listening and speaking, activities 1 and 2**

DON Hello, Don Fisher speaking.
FRAN Hello, Dad, it's Fran.
DON Fran, how are you?
FRAN I'm fine.
DON We're looking forward to seeing you.
FRAN So are we. But I'm ringing to tell you about a change of plan.
DON OK. Go ahead.
FRAN Well, Bruce's friend is staying with us at the moment, so Bruce isn't meeting you at the airport. I'm meeting you.
DON Great!
FRAN And, Dad, sorry about this, but there are only two bedrooms in our house, so you aren't staying with us. You're staying in a hotel near us. Is that ok?
DON Well, ok. But we're still spending a week with you in the city?
FRAN Yes, and then we're driving to Port Stephens, but we aren't spending a week there, we're spending longer- ten days, then we're camping for three days in the outback, not a week. We think a week camping in the outback is too long.
DON That's fine, Fran. Thanks for telling us.
FRAN OK, then, Dad. See you on the twenty-ninth! Love to Mum. Bye.
DON Bye Fran. See you soon.

Lesson 32 Listening, activity 2

JACK Let's go to the cinema.
ANNA I'm sorry but I don't like films.
JACK How about going to the theatre, then?
ANNA Yes, OK, What's on?
JACK An Agatha Christie play.
ANNA Where's it on?
JACK At the Theatre Royal.
ANNA Great!

Lesson 32 Sounds, activity 3

I'm <u>sorry</u> but I don't like <u>music</u>.
I'm <u>sorry</u> but I'm <u>busy</u> tomorrow evening.
I'm <u>sorry</u> but I don't like <u>Spielberg</u>.

Lesson 34 Reading and listening, activity 2

NARRATOR It was ten o'clock in the evening at Ripley Grange. It was quiet ... very quiet. Was there anyone in the dining room? No, there wasn't.
Then, there was a scream. It was Lady Scarlet. Was she in the kitchen? Yes, she was. Was she alone in the kitchen? No, she wasn't. Probe was also in the kitchen. He was dead.
Was there something on the table? Yes, there was – there was a knife on the table. It was red. Was it murder? Who was the murderer?

Lesson 34 Listening and speaking, activity 1

DETECTIVE PRUNE Please answer my questions very carefully. Colonel White, where were you at 8 pm last night?
COLONEL WHITE I was in the dining room. We were all in the dining room, Detective. It was dinner.
DETECTIVE PRUNE Where were you at ten o'clock, Colonel?
COLONEL WHITE I was in the living room. I was with Doctor Plum and Professor Peacock.
DETECTIVE PRUNE Thank you, Colonel. Please ask Miss Green to come in. Miss Green, where were you at ten o'clock last night?
MISS GREEN I was in the garden.
DETECTIVE PRUNE Were you alone?
MISS GREEN No, I was with Mrs Mustard. There was a scream! It was awful.
DETECTIVE PRUNE Thank you, Miss Green. Good morning, Professor Peacock. Where were you at ten o'clock last night?
PROFESSOR PEACOCK I was in the living room with Colonel White and Doctor Plum.
DETECTIVE PRUNE I see. Thank you. Mrs Mustard, where were you last night at ten o'clock?
MRS MUSTARD I was in the garden. It was a lovely evening. I was with Miss Green.
DETECTIVE PRUNE Thank you Mrs Mustard. Doctor Plum, where were you at ten o'clock last night?
DOCTOR PLUM I was in the kitchen.
DETECTIVE PRUNE In the kitchen?
DOCTOR PLUM Yes, Probe was dead. Lady Scarlet was there.
DETECTIVE PRUNE Where were you before you were in the kitchen?
DOCTOR PLUM I was in the living room with Professor Peacock and Colonel White.
DETECTIVE PRUNE Thank you, doctor. And Lady Scarlet, where were you at eight o'clock last night?
LADY SCARLET I was in the dining room with the other people. It was dinner time.
DETECTIVE PRUNE And where were you at ten o'clock?
LADY SCARLET I was in my bedroom. I was hungry and thirsty. There was some food and drink in the kitchen. Probe was there, but he was dead.
DETECTIVE PRUNE So there wasn't anyone in the bedroom with you?
LADY SCARLET Certainly not, constable. How dare you!
DETECTIVE PRUNE I'm a detective, Lady Scarlet. And I know who was Probe's murderer.

Lesson 34 Listening and speaking, activity 5

DETECTIVE PRUNE Thank you for coming to the living room. I now know who was Probe's murderer. You see, Doctor Plum, Colonel White and Professor Peacock were in the living room, and Mrs Mustard and Miss Green were in the garden. Only you, Lady Scarlet, were alone in your bedroom. You were Probe's murderer, weren't you, Lady Scarlet?
LADY SCARLET It's true. I was hungry! I was thirsty! I was bored with cold potatoes, cold meat and tired lettuce! I was tired of water to drink! I was unhappy with Probe.
DETECTIVE PRUNE Thank you, everyone. Lady Scarlet, please come with me.
COLONEL WHITE I say!
DOCTOR PLUM 'Pon my word!
PROFESSOR PEACOCK 'Straordinary.
MRS MUSTARD What's the time?
MISS GREEN One o'clock.
COLONEL WHITE Time for lunch!

Lesson 35 Vocabulary and sounds, activity 3

<u>tel</u>evision <u>vid</u>eo recorder <u>tel</u>ephone <u>rad</u>io <u>com</u>puter <u>per</u>sonal stereo <u>car</u> <u>bi</u>cycle <u>fax</u> machine <u>dish</u>water <u>vac</u>uum cleaner

Lesson 35 Reading and listening, activities 3 and 4

MARY Well, when I was a child, we had a telephone in the hall, but we didn't have a television. No one had television when I was a child. And of course, we didn't have a video recorder.
INTERVIEWER What about a radio?
MARY Yes, we had a radio. The radio was in the living room. The radio programmes were very good.
INTERVIEWER And computers, did you have computers or personal stereos?
MARY Oh, no, we didn't have computers or personal stereos. And I was fifty before I had my own radio. It wasn't a personal stereo.
INTERVIEWER Did you have a car?
MARY Yes, some families had a car, but we didn't. We had bicycles, all six of us.
INTERVIEWER And no fax machine, of course. Or vacuum cleaner or dishwasher?
MARY No, we didn't have a fax machine. But we had a kind of vacuum cleaner, it was a Hoover. And we didn't have a dishwasher. The dishwasher was me!

Lesson 35 Speaking and listening, activity 6

WOMAN Right, let's see how much you know. Did they have aeroplanes in 1950?
MAN Yes, they did.
WOMAN And did they have them in 1850?
MAN No, they didn't.
WOMAN That's correct. How about trains? Did they have them in 1850?
MAN Yes, they did.
WOMAN And in 1750?
MAN No, they didn't.
WOMAN Correct. They had the first train in 1804. And did they have cameras in 1750?
MAN No, they didn't.
WOMAN And in 1850?
MAN Yes, they did.
WOMAN Just, yes. They had the first cameras in around 1830, 1840. Good and newspapers? Did they have newspapers in, let's say, 1550?
MAN Yes, they did.
WOMAN No, they didn't! Did they have newspapers in 1650?
MAN Yes, they did.
WOMAN That's correct. The world's first newspaper was in 1609.
MAN 1609! I didn't know that.
WOMAN Never mind. And the last one - steam engines. Did they have steam engines in, let me see now, in 1750?

MAN Yes, they did.
WOMAN And in 1650?
MAN No, they didn't.
WOMAN Correct. The world's first steam engine was in 1700. Good, well done. Only one wrong answer.

Lesson 36 Sounds, activity 2

lived arrived opened changed listened started hated
liked cooked stopped danced helped

Lesson 36 Listening and writing, activities 1 and 2

INTERVIEWER What did you do for entertainment when you were a child, Ben?
BEN We didn't have television, but we listened to the radio and we played cards a lot. It was fun!
INTERVIEWER And what about sport? Did you do any sport?
BEN I liked football. In fact, I still like football. I listened to the match on the radio. And I played football with friends in the park on Sunday mornings.
INTERVIEWER What else did you do on Sunday?
BEN We often had friends for lunch after I finished the football match. And then we played in the garden in the afternoon.
INTERVIEWER Where did you go on holiday?
BEN Most years we stayed in England. We stayed in a small hotel in Devon. The beach was very close.
INTERVIEWER And what about work. What did your parents do?
BEN My father was an engineer and he travelled a lot. My mother looked after the family. Sometimes I think it was quite hard for her.
INTERVIEWER Judy, tell me about your childhood. What did you do for entertainment at home?
JUDY We liked music very much. We're a very musical family. My brother played the violin and I played the piano. We had family concerts every evening.
INTERVIEWER And sport? Did you like sport?
JUDY Yes, I liked tennis very much. I played with my school friends. In fact I played tennis in my school team.
INTERVIEWER And how did you spend Sunday?
JUDY I stayed in bed until 9 o'clock. Then I did my homework for school. Then my mother cooked lunch for us all, there was always a traditional Sunday lunch at one o'clock. And in the afternoon, we had a walk in the country.
INTERVIEWER Where did you spend your holiday?
JUDY We were very lucky. My parents loved France, and we travelled around France several times, sometimes in the South, sometimes in the mountains, sometimes in Brittany. I still love France.
INTERVIEWER And what did your parents do? What was their job?
JUDY My parents worked in a hospital. My father and my mother were doctors.

Lesson 37 Sounds, activity 2

He <u>didn't</u> live in <u>Barcelona</u> in 1904. He <u>lived</u> in <u>Paris</u>.
He <u>didn't</u> return to <u>Spain</u> to live. He <u>lived</u> in <u>France</u>.
He <u>didn't</u> die in <u>1972</u>. He <u>died</u> in <u>1973</u>.

Lesson 38 Reading and listening, activity 1

STEVE When we started the weather was fine. We walked about fifteen kilometres. About four hours later, we started to come home. Suddenly the weather changed. It was very foggy. We were still in the mountains.
PHILIP Did you stay there?
STEVE No, we didn't. We walked for an hour but it was very cold and dark. We decided to stop walking and wait.
PHILIP Did you wait for a long time?
STEVE Well, no, we didn't. You see, after about two or three minutes, something happened. A dog, a big black dog suddenly appeared out of the fog. He had red eyes. Then he barked at us.
PHILIP Did he want to help you?
STEVE Yes, he did.

PHILIP So we walked for two hours behind the dog. Then we arrived back at the village.
STEVE Did the dog stay with you?
PHILIP No, he didn't. When we arrived at the village, he disappeared into the mountains .
STEVE Did you take a photograph?
PHILIP No, we didn't.

Lesson 39 Vocabulary and sounds, activity 5

drank-drink
spent-spend
got-get
left-leave
flew-fly
saw-see
took-take
ate-eat
sat-sit
went-go
bought-buy
thought-think

Lesson 40 Reading and listening, activity 3

MAN But one day, a strange thing happened. On the night of 28 October 1992 after a cold day in my garden, I had dinner, read my book and then I went to bed. I was asleep when suddenly I heard a noise like a train.
I got up and looked out of his window and saw a train in the garden. "What's happening? Am I dreaming?" I said. There were passengers on the platform.
Some people got out of the train and said hello to their friends, and others got in and said goodbye.
They all wore clothes from thirty years ago. "This is very strange," I thought. I saw an old taxi and two or three old cars, and a man selling newspapers outside the station.
The man stood by a poster. It said, "THE END OF THE WORLD?"

Progress check, lessons 31 to 40 Listening, activity 1

Yesterday, all my troubles seemed so far away,
Now it looks as though they're here to stay,
Oh, I believe in yesterday.
Suddenly, I'm not half the man I used to be,
There's a shadow hanging over me,
Oh yesterday came suddenly.
Why she had to go I don't know
She wouldn't say.
I said something wrong, now I long for yesterday.
Yesterday, love was such an easy game to play,
Now I need a place to hide away
Oh, I believe in yesterday.
Why she had to go I don't know
She wouldn't say.
I said something wrong, now I long for yesterday.
Yesterday, love was such an easy game to play,
Now I need a place to hideaway
Oh, I believe in yesterday.